P

OKANAGAN COLLEGE LIBRARY

P9-BJU-265

GN 281.4 .L54 1991
Uniquely human : the
Lieberman, Philip

146276

DATE DUE

MAR 1 9 1994	
MAR 3 1 1995	
OCT 0 3 2002	

BRODART, INC.   Cat. No. 23-221

Uniquely Human

OKANAGAN UNIVERSITY COLLEGE
LIBRARY
BRITISH COLUMBIA

Uniquely Human

The Evolution of Speech, Thought, and Selfless Behavior

Philip Lieberman

Harvard University Press
Cambridge, Massachusetts, and London, England   1991

Copyright © 1991 by the President and Fellows
   of Harvard College
All rights reserved
Printed in the United States of America
10 9 8 7 6 5 4 3 2 1

This book is printed on acid-free paper, and its binding
materials have been chosen for strength and durability.

Library of Congress Cataloging-in-Publication Data

Lieberman, Philip.
   Uniquely human : the evolution of speech, thought,
and selfless behavior / Philip Lieberman.
      p.   cm.
   Includes bibliographical references and index.
   ISBN 0-674-92182-8
   1. Human evolution.   2. Brain—Evolution.
3. Speech—Social aspects.   4. Behavior evolution.   I. Title.
GN281.4.L54   1991
573.2—dc20   90-38130
                                                                CIP

To Marcia Lieberman

Acknowledgments

I have had a great deal of help from my friends, and thanks are owing to Elaine Anderson, Sheila Blumstein, Ford Ebner, Peter Eimas, Jerry Kagan, Daniel Kempler, Pat Kuhl, Jeff Laitman, Paul MacLean, Jim McIlwain, John Newman, and Walter Quevedo for their comments and advice on the text. Michael Aronson of Harvard University Press likewise suggested many improvements; Matthew Carrano's illustrations clarify many points. Daniel Lieberman has provided invaluable counsel on matters fossil and archaeological. Special thanks are owing to Joseph Friedman, who generously contributed his scarce time to our joint research on how the basal ganglia function to make language and thought possible. I also thank Snow Lion Publications for permission to quote extensively from the work of His Holiness the Fourteenth Dalai Lama, Tenzin Gyatso, *Kindness, Clarity, and Insight* (Ithaca, N.Y., 1984). The John Simon Guggenheim Memorial Foundation has given generous support. Bill Cavness, of WGBH, in a sense started the project. But the strongest claim is Marcia's, who debated these issues with me along the jogging path on Blackstone Boulevard in Providence and on the high trails of the Alps and Himalaya.

Contents

Introduction *1*

1 Brain Structure, Behavior, and Circuitry *11*

2 Human Speech *36*

3 A Thoroughly Modern Human Brain *78*

4 The Brain's Dictionary *112*

5 Learning to Talk and Think *127*

6 Culture and Selfless Behavior *149*

Notes *175*

References *183*

Index *203*

Uniquely Human

Introduction

It was 1966. I was listening to "Reading Aloud" on the Boston public radio station WGBH while soaking in a long, old-fashioned bathtub. The voice on the radio said that chimpanzees cannot talk. "Why not?" I thought.

I spent the next twenty years working on the answer to this question; chimpanzees cannot talk because they lack both the anatomy and the brain mechanisms that are necessary to produce human speech. Human beings possess a unique tongue and mouth that have evolved to facilitate the production of speech; we also possess unique brain mechanisms that allow us automatically to produce complex rule-governed maneuvers of the tongue, lips, lungs, and other organs. These brain mechanisms seem to have evolved in concert with our speech-producing anatomy. But there is more to human language than speech; as the language abilities of chimpanzees and other apes were studied throughout the 1970s and 1980s, it became apparent that apes also cannot form grammatical sentences. Although they can acquire a limited vocabulary by using sign language, they are incapable of grasping the rules of syntax that even three-year-old human children master. Nor do chimpanzees or other animals ever create works of art or complex devices, or convey "creative" thoughts. Nor do they, in their natural state, adhere to the most basic aspects of higher human moral sense. In the chapters that follow, I shall try to demonstrate that human language is a comparatively recent evolutionary innovation that added two powerful devices, speech and syntax, to older communication systems. And I will try to show that the evolution of human speech, complex syntax, creative thought, and some aspects of morality is linked—and that the driving force that produced modern human beings in the last 200,000 years or so was the evolution of speech adapted for rapid communication.

In gaining access to the past I must rely on the only tools available: the archaeological record, the methods of comparative physiology and psychology, and the "synthetic" theory of evolution, which fuses Charles Darwin's insights in *On the Origin of Species* (1859) and modern genetics (Mayr, 1982). I will argue that the principles and mechanisms that Charles Darwin proposed over a century ago account for the evolution of human language and thought. Spots, stripes, and trunks are all "unique" attributes of leopards, tigers, and elephants, but there is nothing unique about the process that led to their evolution. The theory of evolution by means of natural selection always accounts for the special features that distinguish a particular species from others. Few people except creationists would claim that some force outside the ordinary processes of evolution produced the unique teeth, tongue, hands, and various bones of human beings. But the situation is thought to be different for human language and thought—those attributes that, for good or for evil, place us outside the domain of other animals. The origin of language and thought likewise has been traditionally placed outside the range of the natural forces that apply to other animals. The ancient Greeks, for example, attributed language to Prometheus, who also brought fire to man alone. According to *Popul Wuh*, the sacred book of the Maya, "the soothsayers Xpiyaioc and Xmucane were consulted to find out how they could make man." They answered, "You can whittle man and your figures of wood shall speak" (Saravia, 1977, pp. 9–10).

Nature is a miserly opportunist, and the process of evolution always makes use of "old" parts, modifying them to perform new functions. Darwin took account of this when he discussed the evolution of complex, specialized "organs"—"Organs of Extreme Perfection" (1859, pp. 156–194). Human language appears to involve a number of different components, originally adapted to other functions, which have become specialized for language and work together in an integrated manner. Some of the biological mechanisms that underlie human language have a long evolutionary history. All human languages, for example, make use of a melody of speech, an "intonation pattern," that signals the end of a sentence (Lieberman, 1967). The basic intonation, the breath-group, appears to be a modification of the mammalian isolation cry. All mammalian infants use the isolation cry to call their mothers. It is produced by an organ, the larynx, that is similar in humans and rodents (Negus, 1949), and it is controlled by some of the brain mechanisms common to rodents and human beings (Newman, 1990). The brain mechanism that probably underlies our ability to acquire and conceptualize words likewise appears to derive from distributed neural networks, which can be found in many other animals. And many animals, therefore, can comprehend and, in some cases, use words—the most important distinction being the total number of words.

Research over the past fifty years shows that two unique aspects of human language—speech and syntax—enhance the speed of communication. Human speech, as we shall see, has very special properties that allow us to communicate vocally at a rate that is three to ten times faster than we otherwise could. If we lacked human speech we would be limited by the constraints of the mammalian auditory system that make it impossible to keep track of rapid sequences of sounds. The complex syntax of human language likewise overcomes the limits of memory and allows us to keep track of complex relationships between words within the frame of a sentence, again enhancing the speed of communication. However, the evolution of speech and syntax also follows from Darwinian processes. The human tongue, mouth, and brain mechanisms that regulate speech production and syntax evolved from the tongues, mouths, and brains of archaic humanlike animals—hominids who resembled present-day apes in these respects. Organs that were originally designed to facilitate breathing air and swallowing food and water were adapted to produce human speech. Rapid human speech entails the evolution of brain mechanisms that allow the production of the extremely precise complex muscular maneuvers of speech production (Lieberman, 1984, 1985). The brain mechanisms that control speech production probably derive from ones that facilitated precise one-handed manual tasks. Through a series of perhaps chance events they eventually evolved to allow us to learn and use the complex rules that govern the syntax of human language. They may also be a key component of human cognitive ability. Human language is creative; its rule-governed syntax and morphology allow us to form new sentences that describe novel situations or convey novel thoughts. The key to enhanced cognitive ability likewise seems to be our ability to apply previously acquired knowledge and "rules" or principles to new problems.

Thus various parts were added to the brains of the animals who are among our ancestors. The "old" parts were modified for new tasks, and both "old" and "new" parts of the human brain work together to make human language and thought possible, which in turn appear to be necessary elements for altruism. In other words, to paraphrase René Descartes, we are because we talk.

Since much of the debate concerning Darwinism frequently has little to do with his work, it is useful to summarize what Darwin actually proposed.

Genetic Variation and Natural Selection

The physical structure of all living organisms is transmitted by a genetic code. In complex animals, the code is made up of millions of discrete genes, which vary from one individual to the next. Except for identical twins no two individuals of the same species have the same complement of genes; therefore, no two are identical. The results of all studies of variation since Darwin's time show that this is the case for every organism from peas to people.

The reason for the constant presence of genetic variation is the struggle for existence and natural selection. In a sudden insight gained from reading Thomas Malthus' *Essay on the Prin-*

ciple of Population (1798), Darwin realized that all species faced a struggle for existence. Malthus pointed out the problem that still faces developing nations today. As medicine advances and knowledge of public health diffuses, infant mortality decreases and the human population dramatically increases, outstripping available resources of food and land. In the ensuing struggle for existence, people compete for the scarce resources, and only some survive. Others live in abject misery and lead worse lives than they would have if some of the benefits of the modern world had never come into their lives. Darwin realized that the struggle for existence was a general phenomenon, not one limited to humans. In this light the genetic variation that exists in every living species is not a deficit but a mechanism for survival: it allows a species to store adaptations that might prove useful in various contingencies. Natural selection, acting on variation, picks out variations that best fit the present environment, but other genetic possibilities are stored within the pool of genetic variation of the species. Darwin's definition of natural selection is worth repeating:

> Owing to this struggle for life, any variation, however slight and from whatever cause proceeding, if it be in any degree profitable to an individual of any species, in its infinitely complex relations to other organic beings and to external nature, will tend to the preservation of that individual, and will generally be inherited by its offspring. The offspring, also, will thus have a better chance of surviving, for of the many individuals of a species which are periodically born but a small number can survive. I have called this principle, by which each small variation, if useful, is preserved, by the term of Natural Selection. (1859, p. 61)

Natural selection thus acts on *individuals who each vary*. The biological or Darwinian "fitness" of an individual is determined by the degree to which his or her genes are transmitted to successive generations. The presence of constant variation is undisputed (Mayr, 1982). Successful species, ones that adapt to changing conditions, maintain a storehouse of varied traits coded in the genes of the individuals who make up the population that defines the species.

The Mosaic Principle

The "mosaic" principle, in essence, states that living organisms are put together in bits and pieces. The mosaic principle precludes the possibility of everyone's having an identical complex organ such as the "universal grammar" hypothesized by linguistic theorists (Chomsky, 1986). The various parts of the human body are, for example, determined by independent genes. No single gene determines the form of the knee. The upper and lower parts of the knee socket are under independent genetic regulation; people's knees therefore fit together better or worse. The same is true for the finger joints and the parts that must work together in the heart. Although the expression of a gene or group of genes often involves complex interactions with other genes, there are no "master" genes that regulate the fit of the various bits and pieces. Thus if a sudden dramatic change occurs in one gene, the resulting organ will not necessarily fit together functionally with the other body parts, because the genes that regulate these different parts are independent.

If the change involves some critical, basic part of the body, such as the heart or brain, the result usually will not be viable. It might, for example, seem useful to have two hearts. However, an embryo that had two hearts would not survive because the two hearts would have to be connected to arteries and veins whose form is coded by independent genes. The arteries and veins that work with one heart will not work with two—the change is too drastic. In contrast, a mutation that resulted in someone's having a sixth finger would probably be viable because a person could survive whether or not the sixth finger was functional.

Preadaptation

The Darwinian model does *not* preclude abrupt changes; in fact, it accounts for them. Recent theories such as that of punctuated equilibria (Eldridge and Gould, 1972; Gould and Eldridge, 1977) are neo-Darwinian rather than refutations of Darwinian theory. The theory of punctuated equilibria claims that evolutionary changes are not gradual and continuous. Darwin realized that evolution involves discontinuity when he accounted for the presence of "organs of extreme perfection," such as the eye, and for the "transition of organs" in chapter 6 of the *Origin of Species*. Many dramatic changes at first seem inexplicable if evolution by means of natural selection involves only small changes in structure. How, for example, could we account for the appearance of air-breathing animals from aquatic creatures without some sudden major change? There would be no selective advantage for air-breathing lungs in fish. Darwin solved this problem by differentiating between behavior and structure. In discussing the evolution of the lungs, he stated that an "organ might be modified for some other and quite distinct purpose . . . The illustration of the swimbladder in fishes is a good one, because it shows us clearly the highly important fact that an organ originally constructed for one purpose, namely flotation, may be converted into one for a wholly different purpose, namely respiration" (1859, p. 190). Notwithstanding later disputes concerning the preadaptive bases of lungs—some scholars think that swim bladders first evolved from a very primitive type of lungs—the general picture is that some species of fish gradually evolved swim bladders that allowed them to float effortlessly in the water. Some fish, such as sharks, have to swim all the time to keep from sinking to the bottom of the sea. Other fish developed swim bladders, internal elastic sacs that could be filled with air from the gills. These fish were able to float like zeppelins by pumping more, or less, air into the swim bladders. Two swim bladders, positioned like internal water wings, undoubtedly evolved to keep fish from spinning around their long axis. A small change in the structure of some fish, the presence of a slit in the back of the mouth, allowed them to pump air from the atmosphere into their swim bladders and from there into their bloodstream (Negus, 1949). These fish were able to survive when they were stranded in a new environment—dried-up river mudflats. A new, "abrupt" change in behavior, breathing air, resulted from a small change in morphology—a slit connecting the swim bladders to the mouth.

The Darwinian mechanism of preadaptation, in essence, claims that everything evolves from something else.

The preadaptive bases of various specialized organs are sometimes surprising. Milk glands, for example, evolved from sweat glands (Long, 1969). The bones of the mammalian inner ear, which transmit sound to the inner ear and brain, evolved from the joint of the lower jaw. They were still part of the jaw joint in the therapsids, the lizardlike animals that are the ancestors of present-day mammals (Westoll, 1945; MacLean, 1986). The evolutionary history of the human ear explains why people sometimes can develop an earache when they grind their teeth. Preadaptation differentiates between *function* and *structure*. A series of small, gradual *structural* changes can lead to an abrupt change in behavior that opens up a new set of selective forces.

Functional Branch-Points and Sudden Changes

Changes in *behavior* can be abrupt, causing abrupt changes in morphology that we recognize as the formation of a new species. This process again follows from the mosaic principle. Since the different biological bits and pieces that make up a living creature are under independent genetic regulation, the change of one key element can result in an abrupt, qualitative shift in behavior or function (Lieberman, 1984). We can see this phenomenon in the invention of various machines. Thus in the case of the motion picture camera, a new key component allowed evolution toward a different function. Incremental improvements in the technology of producing film, from glass plates, to celluloid sheets, to celluloid rolls, made still cameras more convenient and yielded the *potential* for the development of the motion picture camera when long rolls of flexible film could be manufactured. These improvements in film technology at first had nothing to do with making long rolls of film; rolls of film were introduced because they were more convenient than individual sheets for still photography. But long rolls of film allowed designers to make cameras that could take sixteen pictures per second, which when projected would create the illusion of continuous motion. A *functional branch-point* occurred with the potential for the production of the first motion picture camera. Thereafter the evolution of motion picture cameras diverged from that of still cameras. Still cameras, for example, acquired shutters that can close and open in a short interval to produce sharp pictures of moving objects. Motion picture cameras, in contrast, operate perfectly well with the slow shutter speeds that George Eastman used in his first Kodak cameras. Improvements in the image quality of motion picture cameras involved different factors, such as keeping the frame of film that is being exposed absolutely still while maintaining the movement of exposed and unexposed film. After a century of divergent evolution, still and motion picture cameras are quite different, although they have a common ancestor.

One point that this short discussion reveals is that evolution is not a simple march toward progress. Motion picture cameras are not inherently more "advanced" than still cameras. They have been adapted to different environments and have not replaced still cameras. As a result of continual refinements recent still cameras are in some ways more technically advanced than the newest motion picture cameras—commonly using microcomputer systems to focus and set exposures. Few professional motion picture cameras incorporate these refinements because they are largely irrelevant in the context in which movies are made. The concept of evolution is not equivalent to a linear conception of progress: adaptation occurs in response to different conditions.

The "key" to the evolution of the modern human brain is rapid vocal communication. That consequently is the key to human progress; the enhanced linguistic and cognitive ability that the human brain confers allows us to transcend the constraints of biological evolution. Despite the obvious shortcomings of human behavior there has been slow and uneven progress in the way that people act toward each other and to external nature—in their "moral" sense and conduct. Torture is now considered morally indefensible, so governments that practice torture now hide and deny it. Wanton cruelty toward animals is also usually hidden for the same reason, and the destruction of forests is now "justified," whereas in earlier times the practice appeared to need no defense. However, moral progress occurs precisely *because* it does not follow directly from the mechanisms of biological evolution. Moral progress instead follows from our cognitive ability, which, in turn, derives from our linguistic ability.

1

Brain Structure, Behavior, and Circuitry

Ernst Mayr, one of the major figures in the development of evolutionary biology, has pointed out time and again that the structure and physiology of any living organism necessarily reflect its evolutionary history. The functional organization of the human brain is no exception. The brain appears to work as a complex set of *circuits:* specialized mechanisms with independent evolutionary histories perform different tasks when connected to each other in different circuits. Brain mechanisms that evolved hundreds of millions of years ago work in concert with "newer" parts. And a given part of the brain may participate in several different aspects of behavior that reflect its particular evolutionary history. Moreover, it has become clear that the basic computational elements that constitute the specialized mechanisms of the brain also have a long evolutionary history. Neurons connected together to form *distributed neural networks* can perform different tasks by working in a way that is quite different from the workings of any devices that human beings have hitherto made. Computer-simulated distributed neural networks have been devised that work efficiently as either associative memories or sequential processors. Studies of the brains of simpler animals indicate that they may consist of mechanisms made up of distributed neural networks, which also may form the specialized memories and information-processing devices of the brain circuits that underlie human language and thought.

Modularity

At a 1987 conference of specialists on child development, Jerome Kagan, who has worked for many years on the mental and social growth of children, challenged current theories that claim that the attributes of the human mind derive from isolated discrete mechanisms. He pointed out that we have become trapped by our own words when we study language, thought, social development, and so on as though they were independent entities that had completely independent biological roots. And we have tended to posit separate and distinct "faculties" or "organs" of the brain, each controlling a single and separable aspect of behavior typified by a single word. Kagan also pointed out that we tend to ignore moral sense, which is one of the defining characteristics of human beings. These aspects of human behavior are, to a degree, independent. Some people who have extraordinary linguistic ability may, for example, show little or no moral sense. And we think of linguistic and cognitive ability as being distinct; creative artists, mathematicians, physicists, bankers, and even linguists often cannot express their thoughts. The fact that we use different words to characterize these aspects of behavior has somehow led us to transform them into theories concerning the structure of the brain.

Modular theories such as those of Noam Chomsky and Jerry Fodor (Chomsky, 1980a, 1980b, 1986; Fodor, 1983) embody this error. They claim that the human brain consists of a set of "modules" or "organs," each of which corresponds to and determines one specific behavior. The language organ is supposedly functionally and morphologically independent of other organs that are the basis of thought. In other words, the module or modules that make up the "language organ" do not play a part in any other aspect of human cognition. Chomsky and Fodor differ in this respect. Chomsky claims that the language organ is indivisible, while Fodor (1980) postulates a set of modules: one dedicated to syntax, one to the dictionary of words, and so on, that make up the language "organ." The activity performed in a given module determines a particular mode or component of language or cognition, such as syntax, the linguistic lexicon, speech perception.

Modular theories owe much to phrenology. Phrenology was a respectable science in the early part of the nineteenth century. Franz Gall (1809) proposed the theory, which was modified by Johann Spurzheim (1815). Phrenologists claimed that human behavior consisted of an array of independent "faculties," that each had a "seat" located in a distinct region of the surface of the brain. The surface area of each seat determined the development of some particular faculty, and phrenologists tried to test their theory by measuring skulls. When it became clear that, for example, the seat of veneration was very small in some clerics and large in some murderers, phrenology was discredited.

However, despite the demise of phrenology, modular theories persist. Although they do not claim that the seats of faculties are located on particular parts of the surface of the brain, modular theories of mind retain the phrenological claim for discrete independent devices each of which determines some aspect of human behavior, invoking "logical" design principles to justify their claims. These logical principles presumably lead to the simplest, most logical, and most "economical" description of the neural mechanisms involved in human linguistic ability. However, modular theory is based on the design logic and details of typical digital computers rather than those of biological brains. The hypothetical modules are functionally similar to the "plug-in" components that are often used to assemble complex electronic devices such as digital computers or radars. Modular theories of the brain probably seem reasonable because they correspond with the way that people usually design machines. Modular design is logical when one builds a computer. A system consisting of a set of independent modules, corresponding to the stages of sequential processing that most computer programs use, simplifies the computer's design and facilitates its repair: the servicer can run a diagnostic test, pull out the defective module, and replace it. However, there is no particular reason why the brain of any animal should be designed this way. The brain's computations are not necessarily sequential, and it is impossible to change defective parts. Neuroanatomical and physiological research over the past century shows that the logic of the design of the human brain, like that of any other organ, is a product of evolution.

Models of the brain historically tend to follow the design of the most complex machine of the age. The brain as a telephone exchange was the common model of the 1930s. Digital computers are the present model. The digital computers that have been around for the past thirty years or so are usually constructed of discrete, independent devices that are each "dedicated" to a specific task. But computers do not have to be designed in this way. The circuits could be designed to play a part in more than one function. The printer circuits could also be used to communicate on telephone lines or to display data on the computer screen. Using the same devices for several functions can save parts and conserve electrical power. Modular design was introduced in the early part of World War II in radar sets. The first radar sets were not modular. When they failed, which occurred frequently in the days of vacuum tubes, it was difficult to repair them. Modular design allowed technicians with only basic training to make fast repairs in wartime field conditions. If the designers had instead focused on making radars as small as possible, radar sets might not have become modular. To return to biological terminology, different selective pressures would yield different radar or computer circuitry. We could have a machine that had actual physical "modules" each of which determined some function of the radar set or computer. However, we could also design the machine so that the same devices entered into different behavioral functions. The behavioral functions still could logically be differentiated.

The Circuit Model

Although the past ten years have yielded more information than the previous hundred on how the brain may work, firm facts remain elusive. The following model is a simple, necessarily incomplete attempt to account for some of the phenomena that we know about, and a critique of positions that do not account for present knowledge.

The circuit model of the brain retains one of the basic concepts of the "connectionist" model developed by Norman Geschwind (1965). Geschwind was the guiding spirit of a school of research on aphasia—language deficits produced by brain damage. He demonstrated that many of the language disturbances that had been described since the 1860s could be explained by the hypothesis that different parts of the brain were disconnected when a stroke or accident damaged it—hence the term *connectionist*. Geschwind's observations and theory are still germane. However, like Paul Broca and Carl Wernicke, who pioneered the study of brain and language in the nineteenth century, Geschwind assumed that the brain bases of human language were language-specific and derived from the "newest" part of the human brain, the *neocortex*. The circuit model presented here differs from Geschwind's in that it takes into account two additional factors:

1. Although some brain mechanisms may be language-specific, we cannot assume that *all* the brain mechanisms involved in human language constitute an isolated organ or faculty that is concerned only with language.
2. Although human language and thought probably are the "newest" attributes of *Homo sapiens*, their brain bases are not restricted to the phylogenetically newest parts of the brain.

Thus the circuit model takes into account the evolutionary history of the human brain. We undoubtedly have specialized neural organs, mechanisms that evolved to facilitate various "higher" cognitive and linguistic activities. However, they evolved from mechanisms that worked to make "simpler" activities possible. And they usually continue to participate in the older, simpler pattern of behavior as well as in the newer, derived cognitive activity. As Darwin noted in 1859, organs that are specialized for "new" functions always evolve from organs that had some other function. The new behavior almost never completely displaces the old activity. For example, the human larynx, which has become adapted to produce sound efficiently, evolved from a simple valve that could abruptly close to protect the lungs. The human larynx retains this primitive function.

Similarly, the limbic system of the brain consists of a set of parts that are involved not only in the control of such basic functions as the regulation of the autonomic nervous system (such as temperature) and the control of skeletal muscle, but also in memory and the perception of pain and pleasure. Like other specialized organs, the brain mechanisms that make human language and thought possible probably also continue to participate in older, simpler activities. Pervasive extreme modularization is most unlikely, although some faculties such as vision involve very specialized brain mechanisms that do not enter into other aspects of behavior. In fact even vision involves the interaction of parts of the brain that function in other aspects of behavior.

Comparative Neurophysiology

Paul MacLean's comparative studies (1967, 1973, 1985, 1986) of the brains of humans and simpler animals provide a good starting point. Figure 1–1 shows the three main components of the human brain. The brain can be thought of as resembling a peculiar sort of sprouted onion, consisting of a few extremely thick layers contained inside an extremely hard skin—the skull. The main body of the onion inside the skull is the cerebrum; the midbrain and brainstem form the sprouted stem. The innermost part of the cerebrum contains a number of structures that anatomists call the *basal ganglia*. Although the basal ganglia of human beings are more differentiated and proportionately larger (in relation to body weight) than those of simpler mammals such as rodents or reptiles (Parent, 1986), they derive from the brains found in reptiles. MacLean therefore calls this part of the brain the reptilian complex. The basal ganglia are connected to the spinal cord by the midbrain. However, certain parts of the midbrain, such as the *substantia nigra*, are so closely

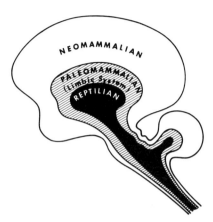

Figure 1-1. Diagram of the "triune" evolution of the mammalian brain. Comparative studies, together with an examination of the fossil record, indicate that the mammalian brain retains the anatomical and chemical features of three basic formations that reflect an ancestral relationship to reptiles, early mammals, and late mammals. The neomammalian component includes the neocortex; the paleomammalian includes the cingulate cortex; the reptilian includes the basal ganglia and midbrain structures. (After MacLean, 1985)

connected to the basal ganglia that they are considered to be associated structures that are functionally part of the basal ganglia circuits (Parent, 1986).

The basal ganglia in humans have traditionally been associated with the control of motor activity. Parkinson's and other neurodegenerative diseases that affect this part of the brain cause deficits in motor control—the victim's gait and hand movements often become labored and distorted, and tremors typically occur. However, recent data show that damage to the basal ganglia also produces speech, language, and cognitive deficits. The basal ganglia are biochemically and anatomically different from the next layer, the *cingulate cortex* or old motor cortex. For example, the cells that make up the brain—the neurons—depend on various chemical *neurotransmitters* that are a basic part of the process whereby electrical signals are generated and transmitted from one cell to another. Whereas the neurotransmitter dopamine is necessary to maintain normal functioning of the basal ganglia, other neurotransmitters regulate activity in the cingulate cortex.

Reptiles to Mammals

The cingulate cortex first appeared in simple mammals. It is involved in much more than simple motor control; in human beings and other mammals it regulates smell and various aspects of emotion and general behavior, such as aggression and mood (Mesulam, 1985). Studies of the effects of brain lesions on behavior show that it also regulates mother-infant interaction and the vocalizations that infants use to summon aid (Newman and MacLean, 1982; Newman, 1985; Sutton and Jurgens, 1988).

Mammals evolved from the therapsids, mammal-like reptiles who lived long before the dinosaurs, between 230 and 180 million years ago. Therapsid fossils have been found on every continent (including Antarctica) because the world then consisted of one large continent—the great land mass of Pangaea.

MacLean (1986) has pointed out three clear changes in behavior marking the evolutionary transition from reptiles to mammals: (1) mammals nurse their infants; (2) infant mammals have to maintain contact with their mothers and therefore produce *separation* or *isolation calls* (the terms are used interchangeably) when they are separated from their mother; (3) mammals play. Experiments that destroy part of the brain in mammals such as rats show that lesions in the cingulate cortex result in marked deficits in maternal behavior (Stamm, 1955; Slotnick, 1967). The rats neglected nest building and nursing and retrieving pups; only 12 percent of the pups survived.

The separation calls of mammals are obviously an essential part of the evolutionary discontinuity that differentiates mammals from reptiles, and it therefore is not surprising to find that the separation calls of other mammals also are regulated by the cingulate cortex. A series of experiments that MacLean started in 1945 shows that the anterior (front) part of the cingulate cortex elicits the separation call in the monkey. Figure 1–2 shows the similarities between the separation calls of a squirrel monkey, a macaque monkey, and a human. The normal crying pattern of human infants conforms to the general primate form (Lieberman et al., 1972; Newman, 1985). The pitch of the infant's voice first rises, stays almost level, and then falls. We

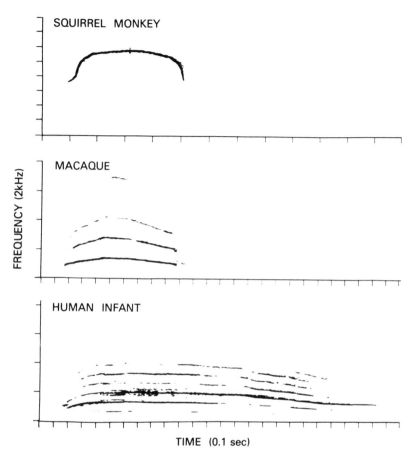

Figure 1–2. Spectrograms showing the "intonation" (basic melody) of the separation calls of a squirrel monkey, a macaque monkey, and a human infant.
The horizontal axes show time. The vertical axes show the pitch of the voice
in the form of the fundamental frequency of phonation, F0, and some of its
harmonics. (The harmonics occur at multiples of F0—for example, 2F0, 3F0,
4F0—and essentially magnify the pattern of variation.) The pattern of fundamental frequency variation is similar for these primates over a complete expiration despite different average pitches and durations. (After MacLean, 1985)

retain this vocalization pattern, although we have adapted it for
language, using it to segment the flow of speech into sentences
(Lieberman, 1967).

Several anatomical attributes work with the brain to achieve
the types of behavior that differentiate mammals from reptiles.

If vocal communication is to be effective in maintaining the mother-infant bond, the mothers obviously have to be able to hear their infants' separation calls. Therefore, unlike reptiles, mammals have a middle ear, which allows them to hear quieter sounds. The middle ear has two small bones, the *malleus* and *incus* (hammer and anvil), which evolved from two small bones of the joint that hinges the jaw to the skull. These bones, which were still part of the jaw hinge in therapsids, act as a sort of mechanical amplifier. The first true mammals can be identified in the fossil record because they had middle ears. It is difficult to make any statements concerning the organization of an animal's brain from the remains of its skull. However, the bones of the middle ear are an index for the presence of a cingulate cortex in these fossil mammals. The specialized anatomy that accompanies the specialized brain mechanism in living animals allow us to make other reasonable inferences concerning the fossil record. In the following chapters we will use correspondences between the bones that signal specialized speech-producing anatomy and the brain mechanisms that regulate human speech to interpret the fossil record of human evolution.

Therapsids may have had other mammalian physical attributes. They probably had milk glands, which derive from sweat glands. They also may have had muscular lips, which allow the young to suckle. Their mouths also had a secondary hard palate (discussed in the next chapter), which allowed them to chew pieces of food into small pieces and breathe at the same time. This behavior also differentiates mammals from reptiles. The evolution of mammals thus appears to derive from a change in behavior—nursing that resulted in the "perfection" of brain mechanisms and anatomy that enhanced successful nursing, thereby increasing biological fitness.

"Higher" Mammals

The evolutionary transition to higher mammals involved the addition of the neocortex to the older brain. Harry Jerison's (1973) comprehensive study of the evolution of the brain shows that the size of the brain in relation to the total size of an animal is one of the best indexes of the relative intelligence of a species.

The increased relative size of the brain in higher mammals reflects an increase in the size of the neocortex (Brodmann, 1908, 1909, 1912; Jerison, 1973). The neocortex enhances the performance of complex motor tasks, the perception and interpretation of sensory information, and the integration of the senses with thought and action. It appears to be the part of the brain that is most involved in thinking—in being able to respond to new situations and learning new responses (Stuss and Benson, 1986).

Studies comparing the brains of related species permit reasonable inferences about its evolution. Humans differ from all other mammals with respect to both the relative size of their brains (Jerison, 1973)[1] and the proportions of the different parts of the neocortex. The frontal regions of the neocortex are much larger in humans than in other primates (see Figure 1–3). There is also a qualitative difference in functional organization: in humans the neocortex is involved in the voluntary control of speech. This is not so for other primates. In monkeys, for example, electrical stimulation of the neocortex does not affect vocalization. Instead, stimulation of the anterior cingulate cortex elicits vocalizations such as the isolation call that have a volitional component; stimulation of various parts of the midbrain elicits cries that are part of their fixed range of emotional displays (Sutton and Jurgens, 1988).[2]

The voluntary control of vocalization in human beings appears to be another example of evolutionary "add-ons" and concurrent modifications of older coordinate structures; the modifications probably involve the neocortex, the cingulate cortex, the basal ganglia, and some of the pathways connecting these structures to the muscles of the lungs, larynx, upper respiratory system, and mouth. Coordinate changes also occurred in the anatomy of the mouth, tongue, and throat. These changes cannot be viewed as the simple replacement of old parts with new ones that have no analogue in the animals to which we are related. This is patently obvious for the anatomy that we can see and touch. Chimpanzees, dogs, and cats have tongues, jaws, and mouths. It is also true for the brain. The mammalian brain did not lose the reptilian complex. Mammals added the cingulate cortex, which adds some other behaviors and enhances motor control, but the older part of the brain was not unplugged and replaced by a new "module." It continued to function in concert with the newer part. Nor did primates trade in the cingulate cortex for the new improved neocortex.

The older parts of the brain also have evolved and changed in more advanced animals in response to selective pressures from new functions and the newer parts of the brain. Although most comparative anatomists stress the enlargement of the newest part, the neocortex, a similar change occurred in the human basal ganglia: the *caudate nucleus* and *putamen* (discussed in more detail in Chapter 3) are fourteen times larger in humans than they would be if we had the brain of an insectivore, a primitive primate with the same body weight. The basal ganglia have also become more complex and differentiated. In rodents, who are similar in many ways to the first mammals, the putamen and caudate nucleus form a single entity. In higher primates such as squirrel monkeys and human beings they have become differentiated and play a part in different aspects of behavior (Parent, 1986).[3]

Many of the gestures and postures that higher primates such as squirrel monkeys use for communication are still controlled by the basal ganglia; these displays are analogous to displays that lizards use to challenge competitors. In lizards and squirrel monkeys partial destruction of the basal ganglia abolishes the territorial displays to challenge competing males (MacLean, 1985, 1986). The basal ganglia also appear to be implicated in many of the nonvocal displays that signal or accompany emotion and to play a role in integrating the performance of the activities that make up an animal's daily routine. In other words, many aspects of mammalian daily behavior continue to be controlled by the reptilian complex. This holds for all mammals, including humans (Mesulam, 1985). We will see in Chapter 3 that the basal ganglia are also implicated in human language and thought.

Cortical Organization and Circuitry

Although our understanding of the functional organization of the cortex (the term embraces both the neocortex and the cingu-

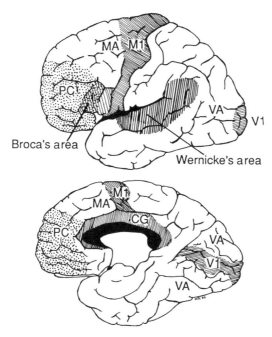

Figure 1–3. Lateral and medial views of some of the functional areas of the human brain. The regions that appear to be involved in speech, language, and some aspects of cognition include Broca's area, Wernicke's area, the prefrontal cortex (PC), the motor association area (MA), the primary motor area (M1), and cingulate cortex (CG). V1 and VA are visual primary and association areas.

late cortex) is far from complete, a great deal is known. Figure 1–3 is typical of illustrations showing the functional organization of the human cerebellum. The upper diagram is a side view of the exterior left surface, with the prefrontal region, PC, at the left. The lower diagram shows a side view of the brain sectioned through on its midplane (analogous to a view of the inside of a grapefruit cut in half), with the cingulate cortex, CG, toward the interior, under the neocortex.

The phrenologists were not completely wrong in localizing brain functions. Some parts of the cortex appear to be extremely specialized and modality-specific. Those that are most closely tied to specific functions are the ones that have the closest connections to the senses. Area V1 in Figure 1–3, for example, is a "primary visual" area; it is essential for vision. Cortical areas that are somewhat removed from the connections from the eye, such as a "visual association" area, VA, are modality-specific association areas. Damage to these association areas gives rise to complex visual perceptual deficits. For example, people afflicted with such damage may be unable to recognize faces although they can see and recognize other shapes. Humans, like other primates, have very extensive visual association areas. Motor control is also divided into "primary areas," M1, and "premotor" and motor association areas, MA. The primary motor areas directly control the activity of particular muscles. Electrical stimulation of the appropriate primary motor area in a monkey, for example, will result in the motion of its lips (Sutton and Jurgens, 1988). The motor association area of the cortex is involved in the control of complex motor movements. Damage to motor association areas in monkeys results, among other things, in impairment of motor tasks that involve visual guidance.

Broca's area, the part of the cortex that Paul Broca in 1861 associated with the control of human speech, has no functional equivalent in nonhumans. Permanent language deficits can result when extensive brain damage cuts the connections to this area. The prefrontal cortex is an association area that receives information from all over the brain. Damage to the prefrontal cortex in humans and other primates results in a loss of "vigilance," the ability to pay attention to a problem, as well as in several other deficits in learning and performing complex tasks, and abstract thought. Other parts of the cortex and the subcortical parts of the brain also appear to play a part in human language and cognition, but a great deal of uncertainty and conjecture remains concerning the functions of most parts of the brain (see Mesulam, 1985, and Caplan, 1987, for comprehensive reviews).

However, although the functions of the various parts of the cortex are not certain, they have somewhat different structural properties, and standardized systems have been devised to identify various parts. The system that is most often used was introduced by Korbinian Brodmann (1908, 1909, 1912), who studied the "cytoarchitectonic" structure of the various parts of the brains of humans and other animals. The microscopic cytoarchitectonic structure of different parts of the brain differs. For example, the proportions of various types of neurons in the parts of the cortex that are known to be involved in motor control differ from those of prefrontal regions. Comparative cytoarchitectonic studies therefore allow us to compare the proportions of different parts of the brain in different species. The evolution of some aspect of the brain can be studied in this way by charting the proportions of particular structures in simpler and more complex related species. Figure 1–4 shows Brodmann's sketches of the surfaces of the brains of two phylogenetically lower and higher primates. The sketches show the left hemisphere of the brain; the anterior part of the brain is to the left. Various areas are labeled with their Brodmann numbers.

Brodmann was working in a "locationist" tradition that derived from the phrenologists. He divided the various parts of the cortex into areas that he believed were specialized for various functions. Area 4, the *primary motor area*, for example, directly controls the individual movements of various parts of the body. It receives inputs from the *premotor area*, which includes areas 6, 8, 43, 44, and 45. The premotor area stores patterns of learned motor activity. Broca's area consists of Brodmann areas 44 and 45; it is usually classified as a premotor area by virtue of its cytoarchitectonic structure and its involvement in the motor control programs that produce human speech. The prefrontal cortex lies anterior to the premotor areas and includes areas 9-15, 46, and 47 (Stuss and Benson, 1986).[4] However, virtually all of these areas are connected to each other by complex pathways that can involve the basal ganglia and other older parts of the brain.

Note the increase in the relative size of the frontal regions of the brain shown in Figure 1–4 when we compare macaque monkeys to human beings. The human brain does not simply have a larger neocortex than a monkey or ape brain; our prefrontal cortex is proportionately larger. Terry Deacon, who has been studying the brains of many intelligent mammals—including monkeys, apes, and whales—notes that the human prefrontal cortex is twice as large as it would be if we simply had a large ape brain (Deacon, 1988b).

The connection between the neocortex and language has

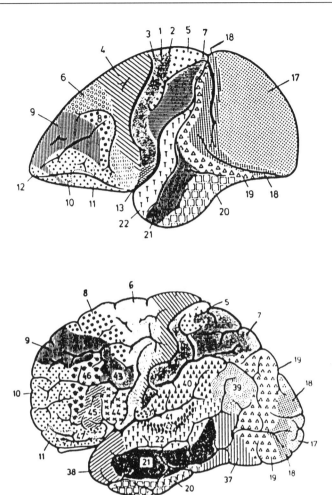

Figure 1–4. Brodmann's (1909) cytoarchitectonic maps of the macaque monkey (top) and human (bottom) brains. The frontal regions of the neocortex appear as the upper leftmost areas. The prefrontal cortex, which consists of areas 9–15, 46, and 47, is proportionately larger in the human brain.

been evident since Broca noted that massive lesions in and near this area produced deficits in speech production. Broca's patient could still execute the gestures that enter into speech production, but could not put them together in the complex coordinated articulatory maneuvers necessary for speech. It is also worth noting that the patient had also lost the use of his right arm—deficits in motor control outside the domain of speech often occur with speech production problems (Liepmann, 1908; Kimura, Battison, and Lubert, 1976). Victims of Broca's aphasia appear to lose control of some part of the set of instructions, the motor "program" that specifies the production of speech. They can, for example, close their lips, but they cannot coordinate the controlled lip-closing movement necessary to produce the [b] of the word *bad* with the activity of the larynx. Recent data (discussed in Chapter 3) show that Broca's aphasia also involves deficits in the comprehension of syntax. In contrast, lesions in or near an area of the cortex close to the auditory association area (AA in Figure 1–3) result in Wernicke's aphasia. The patient is unable to comprehend spoken language. Speech production is not impaired except that inappropriate words may be used.

Other types of aphasia can be explained by means of the connectionist theory developed by Norman Geschwind (1965)—that is, that particular language deficits derive from lesions that interrupt *circuits* (Benson and Geschwind, 1985). For example, lesions in the connection between Broca's and Wernicke's areas, the *arcuate fasciculus,* result in conduction aphasia, characterized by sharp differences in speech comprehension and production: patients may insist that they know the correct name for an object and be able to comprehend its meaning, but be unable to produce the word spontaneously. However, it has become clear that the circuits are far more complex than was thought and that they involve the "old," subcortical parts of the brain as well as the neocortex (Stuss and Benson, 1986; Alexander, Naeser, and Palumbo, 1987). Moreover, the behavioral deficits of victims of aphasia are not restricted to language. Motor control and cognition (Stuss and Benson, 1986; Kimura, 1988) are also impaired.

Brain Circuits and Behavior

Brodmann's charts showing labeled cortical areas fit into the context of the "locationist" model of the brain that was current at the start of the century. Broca likewise believed that he had found a specific location in the human brain that regulated speech. Sigmund Exner (1881) continued along these lines and claimed that various complex behavioral functions are each controlled by a specific and dedicated center in the brain. However, these archaic theories are wrong. As M.-Marsel Mesulam notes in his review of the theory and practice of neurology, theories such as Exner's are "inconsistent with much of the evidence . . . that shows the interdependence among cerebral areas in all phases of information processing" (1985, p. 57). Mesulam, in a wide-ranging discussion of neurological deficits, concludes that a circuit model best explains the way that the brain works. He notes that:

1. Complex functions are represented within . . . interconnected sites which collectively constitute a [circuit].
2. Each individual cerebral area contains the neural substrate for several sets of behaviors and may therefore belong to several [circuits].
3. Lesions confined to a single region are likely to result in multiple deficits.
4. Different aspects of the same complex function may be impaired as a consequence of damage to one of several cortical areas or to their interconnections. (1985, p. 58)

Mesulam's view is consistent with most studies of the workings of the brain. Different circuits connect the various parts of the brain in different configurations to accomplish different tasks. The complexity of the different circuits reflects the historical logic of evolution—different behaviors could be accomplished at minimum biological cost by adding a single "new" component, making use of "older" brain mechanisms. The neurophysiologist Jean-Pierre Changeux, for example, notes that "apparently simple operations of behavior, such as the movements of the eye or the killing of a mouse by a cat, in fact involve the recruitment of a large number of neurons (thousands, even millions) from many different areas of the brain . . . A given behavioral act may engage, *simultaneously* and *necessarily*, groups of neurons which appeared at different periods in the evolution of vertebrates" (1980, p. 188). Killing a mouse may not seem like a very "high" cognitive act, but hunting is no simple task. It involves skill and cleverness; the cat must be able to outwit the mouse. (This is not as important for domesticated cats, for whom cleverness is not as crucial; they do not have to work very hard. Indeed, it is the case that domesticated animals have smaller brains than their wild brethren; Kruska, 1988.)

A simple experiment demonstrating vision-stabilizing circuits. Even "simple" tasks, such as looking at this book while you continually move your eyes, require the integrated activity of different parts of your brain. A circuit involving the parts of the brain that control eye, head, and body movement, the frontal regions of the brain, and the posterior parts of the brain directly involved in vision corrects for the constant motion of your eyes. Consider the bane of home movies and videos—the shifting, lurching, dizzy world that results from camera movement. Novice camera operators do not realize that they have to stabilize the camera, because the world does not suddenly move when you move your head or eye. What is not intuitively obvious is that the stability of the perceived visual world derives from an elaborate process in which the primary visual areas of the brain receive information from "higher" areas of the brain. The reason that the world doesn't appear to move is that the human visual-sensory system has been prepared to compensate for change in position by knowledge of the motoric instructions to the eyeball, neck, and the rest of your body. The motor control systems that control the direction in which you are looking and eyeball movement speak to the brain mechanisms that interpret visual images. The frontal areas of the brain that integrate motoric and sensory information form part of a circuit that is necessary to interpret the basic visual input (Teuber, 1964).

You can easily, literally "see" that the brain network that

stabilizes vision really exists. Simply position your right index finger against the right side of your right eye and gently push it inward. The reason the image jumps is probably that there was no selective advantage for any activity that involved people's pushing against their eyes. Therefore, natural selection for the integration of eye movement from finger pushing never occurred.

Distributed Neural Networks

In the past ten years, numerous scientific conferences have focused on the properties of distributed neural networks (for example, Rumelhart et al., 1986; Edelman, 1987; North, 1987; Anderson, 1988). Distributed neural networks are a reasonable model for the basic computational architecture that makes up the different specialized mechanisms of the brain. The relationship that holds between these networks and the specialized parts of the brain connected in circuits is similar to that which holds between transistors and electronic devices. Transistors are the building blocks from which electronic devices can be constructed. The same transistors (or integrated circuits that consist of many transistors) can be connected in different ways to fashion a radio or a digital computer. Distributed neural networks likewise can be assembled in different configurations to function as extremely rapid sequential processors or as memory devices.

Although distributed neural networks are usually simulated on digital computers, they derive from biological thinking. In 1949, before any digital computers were generally available, Donald Hebb proposed that brains were distributed networks in which memories were stored by modifying the *synapses,* the electrical connections between the neurons that make up the brain. Hebb claimed that learning will take place when two neurons, each of which responds to a particular stimulus, are simultaneously activated, causing the synapse between them to be strengthened. The modified synapse links the two neurons into a circuit; the circuits form the memory traces. Some psychologists immediately realized that the Hebbian model

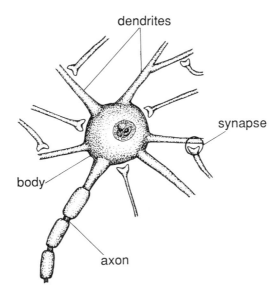

dendrites

synapse

body

axon

Figure 1–5. Sketch of a neuron showing axon, dendrites, and some synaptic connections.

provided a biological basis for the process of associative learning.

The elementary units that make up the nervous system of all animals are neurons. Each neuron is a cell *body* that can generate a sharp, abrupt *pulsatile* electrical signal, which can travel outward on a pathway called the axon. Each neuron also receives signals from other neurons on pathways called dendrites. Figure 1–5 shows a sketch of a neuron with its axon, dendrites, and some of the synapses. The process by which neurons communicate with other neurons involves a gradually increasing signal coming from a dendrite being sensed by the cell body, which then generates a pulsatile electrical signal that is transmitted outward on its axon to other neurons. The connection between each axon and another dendrite or cell body is made through a *synapse*. Each synapse acts as a sort of amplifying or inhibiting relay station, adjusting the degree to which the cell body will generate an output signal in response to the incoming signals. Imagine a relay race in which a verbal message must be passed along in order to initiate the next stage of the race, but all the runners are hard of hearing and have hearing aids. The volume control of each hearing aid can be adjusted to amplify sounds or not. The hearing aid can be regarded as a synapse. It does not directly regulate a runner's legs, but if the volume is turned down the runner will not hear the message; turning up the volume of the hearing aids will increase the "sensitivity" of the system and cause more running.

Electrical Power Networks: A Distributed System

The distributed neural networks that make up the brain consist of millions of interconnected neurons. The key to their operation is the connectivity, but it is difficult to describe how they work because very few concrete examples of simple man-made distributed systems exist. Although practically all mechanical or electronic devices store information, they usually store the information in a discrete—that is, a specific—location. The heating system in a house stores the desired temperature in the thermostat. A microwave oven has an electronic clock mechanism in which the time that the oven will cook is stored. However, distributed systems do exist in the electric power grids that link the generating stations and users of electricity in most developed countries.

The first electric systems were simple, discrete systems. Generators were built in densely populated neighborhoods, and wires were strung to the homes and businesses of the subscribers. The way in which the customers were connected to the generator was not fundamentally different from any other system of production and consuption that makes use of discrete components. A break in the wires that connected the generator to the customer resulted in loss of service to the customer. If the generator broke down, the supply of electricity was interrupted for everyone. Electric companies modified their distribution systems to avoid these problems. As more electricity systems were built it became possible to connect generators to one another. Power could be transferred from remote generators, ensuring service even if the local generator failed. Extra, "redundant" transmission lines were gradually added to guard against failure of a particular line. It became possible to connect many generators and customers into networks covering thousands of miles. Although one generator may fail, the flow of electricity will not fail *if* the other generators have extra power-generating capacity that in normal circumstances would be redundant. Through a complex interconnected network of generators and "switching centers" the system adjusts and redirects power from other generators, and apportions output to different users.

The property that makes an electric power grid a distributed system is the interconnection of switching centers. Although the failure of a single generator could be offset by simply diverting power from another generator, that usually does not occur. The power grid usually compensates by diverting power from many different generators; the switching centers throughout the entire grid can readjust by shifting a little power from each generator. The consequence of the generator failure thus would not be reflected in a single *discrete* location. The electrical representation of this and other events is also distributed throughout the system and is represented by the values that the switching points have assumed. In other words, the settings of the switching points throughout the system constitute the "record," the "memory trace," of each event. Memory resides in the switch settings throughout the network.

Computer-Implemented Studies

The distributed neural networks studied so far have been extremely simplified computer models of the presumed microstructure of the brain.[5] Nonetheless they have a number of features in common with real brains. First, they are resistant to local disruption—destruction of part of a distributed neural network slows it down and makes its decisions less perfect but does not totally stop it from functioning. Karl Lashley (1950), for example, removed 90 percent of the visual cortex of cats without seriously impairing their discrimination of visual forms. Similarly, the studies that we will review in Chapter 3 show that minor damage localized to Broca's or Wernicke's or other areas of the brain does not result in permanent language deficits. The victim usually recovers after a few months. Distributed neural networks have a second brainlike cognitive property: they inherently set up patterns of association and "learn" as they are exposed to different events or information. They offer an explanation for the way that animals can quickly learn to abstract a general "principle" from a limited number of exposures to a set of items or events that exemplify a general concept. Pigeons, for example, can learn to identify trees or species of fish after fairly short training sessions in which they view color slides of different individual trees or fish (Herrnstein, 1979; Herrnstein and de Villiers, 1980). The trained pigeons do not simply recognize the particular images that have been presented in the training sessions; instead they recognize similar trees or fish presented in new color slides. The pigeons have learned to abstract the concept that underlies the varying instantiations. Distributed neural networks behave in much the same way. Computer-simulated distributed neural networks have been used to learn syntactic "rules" (Schrier, 1977; F. Z. Liberman, 1979) or to learn to identify pictures of dogs as being dogs by extracting a prototype construct of dogginess after viewing a large number of pictures of dogs and other quadrupeds. They have been used to recognize photographs of human faces; after being trained by "seeing" 500 faces the distributed neural network could recognize them on viewing fragments of the photographs (Kohonen, 1984). Tauvo Kohonen and his associates have also used a distributed network to recognize words. Kohonen's system works more slowly and renders somewhat fuzzy images of faces after half of its computing elements are disconnected, but it nonetheless works.

Distributed neural networks are biologically plausible mechanisms for this aspect of cognition—the ability to abstract a general principle or concept from a set of particular examples each of which differs slightly. As Alun Anderson (1988), commenting in the "News and Views" section of the journal *Nature*, notes, "they make complex phenomena emerge from simple principles in a way that seems reassuringly biological" (p. 65). The neural networks that were discussed at the conference that Anderson is reporting were used by Terrance Sejnowski and his colleagues to simulate the cortical visual "maps" that occur in the brain (Sejnowski, Koch, and Churchland, 1988). When it is exposed to the light patterns that have been used for decades in research on the visual system of cats, the network acquired properties like those of the cat's primary visual cortex cells (Hubel and Weisel, 1962). Because the cat's primary visual cortex responds to bars and edges in a very specific way, many people had thought that the very specific responses were innate characteristics of the organism. However, Sejnowski's network quickly made these same specific responses *without* prior—that is, innate—detailed information. Work with distributed neural networks shows that we have short-changed the role of learning. Distributed neural networks appear to be able to learn, store, and access information about the environment. They are reasonable models for the hardware that makes up the dictionary that is in our brain—an "active" dictionary that learns the meanings of words through real-life experiences.

2

Human Speech

Roman Jakobson's accomplishments span many fields—linguistics, literary criticism, semiotics, structural anthropology, mushroom collecting—but his most influential contribution to science was to bring nativist linguistic theory to the United States. Jakobson fled from Russia to Vienna as a result of the Bolshevik revolution. During World War II he fled again to Sweden and from there to America, where he ultimately held a joint professorship at Harvard University and the Massachusetts Institute of Technology. Although current nativist linguistic theory focuses on syntax, its strongest claim—that innate brain mechanisms determine the particular form of any human language—derives from Jakobson's work on the sound pattern of language. Moreover, the strongest evidence for innate linguistic brain mechanisms comes from the study of speech. These data are generally consistent with the theory that Jakobson brought to the United States.

American linguists in the 1930s were not concerned with the universals of language; instead they stressed diversity, which presumably reflected differences in culture. In contrast, Jakobson looked for similarities that had a biological basis. Both linguistic traditions probably derive from a common historical factor—territorial expansion. One aspect of the opening of the American West and the acquisition of the eastern and southern provinces of imperial Russia is similar: in both instances exotic native cultures and languages were encountered. However, whereas American linguists focused on the differences between the languages of the native inhabitants of North America, Jakobson saw a set of common universal elements in the exotic languages of the Uzbek, Kabardians, and other "nationalities" that still form the Soviet Union.

According to Jakobson's theory, the sounds of all human languages are composed of atomic units, called "features," and all humans innately possess the biological bases of these features. Individual languages use subsets of these features. Normal children can learn any language because they have innate knowledge of the entire set. Language acquisition thus involves a sort of activation of the particular features that a given language uses; as people mature, they lose the unused ones. Although not all aspects of speech have an innate basis, the research of the past thirty or so years is generally consistent with this theory. It has become evident that human speech involves a number of innate, genetically transmitted anatomical and neural mechanisms. These biological attributes appear to differentiate anatomically modern humans from all other living animals as well as from comparatively recent fossil hominids such as the Neanderthals, who disappeared less than 35,000 years ago. The concurrent evolution of modern human beings and of human speech is the focus of this chapter.

The Sounds of Speech

Until the 1960s it was not realized that human speech is itself an important component of human linguistic ability. Linguists thought that any set of arbitrary sounds would suffice to transmit words. Research efforts to design a machine that would read books to blind people demonstrated that the sounds of speech have a special status (Liberman et al., 1967). Speech allows us to transmit *phonetic segments* (approximated by the letters of the alphabet) at an extremely rapid rate—from fifteen to twenty-five per second. This fact leads to a seeming mystery. George Miller, one of the founders of cognitive science, in 1956 published a paper called "The Magical Number Seven, plus or minus Two: Some Limits on Our Capacity for Processing Information." Miller showed that humans cannot identify nonspeech sounds at rates that exceed seven to nine items per second. We do no better when we process visual information. How, then, can we possibly understand speech, which is typically transmitted at a rate of about fifteen to twenty-five sounds per second? The answer is that specialized anatomy and brain mechanisms allow us to make these speech sounds, and a set of brain mechanisms allows us to "decode" speech signals in a very special way.

For example, a short sentence such as this one contains about fifty speech sounds. All fifty can be uttered in two seconds, and human listeners have no particular difficulty understanding what has been spoken. Transmitted at the nonspeech rate, the sentence would take so long that a listener might well forget the beginning before hearing its end. The engineers working on the reading machine discovered that only speech sounds would allow people to understand the meaning of even moderately complex sentences. Other sounds, such as Morse code, were slow and required the listener's full attention at the expense of content. At a rate of fifty words per minute, equivalent to about five sounds per second, a Morse code expert is too busy transcribing the code to comprehend the message's meaning.

The high transmission rate of speech is thus a crucial component of human linguistic ability, because it allows complex thoughts to be transmitted within the constraints of short-term memory. Although sign language can also achieve a high transmission rate, the signer's hands cannot be used for other tasks. Nor can viewers see the signer's hands except under restricted conditions. Visual hand signs still function as part of the linguistic code (MacNeill, 1985), but the primary linguistic channel is vocal. Vocal language represents the continuation of the evolutionary trend toward freeing the hands for carrying and tool use that started with upright bipedal hominid locomotion. The contribution to biological fitness is obvious. The close relations of hominids who could rapidly shout "There is a lion behind the rock!" were more likely to survive, as were hominids who could convey the principles of toolmaking in comprehensible sentences. Human speech also has some lesser selective advantages; the sounds that are specific to speech—that is, those that only humans can produce—are less susceptible to perceptual confusion than the sounds that other primates can make. These perceptual factors, discussed below, may have had a primary adaptive role in the initial stages of the evolution of human speech.

The Physiology of Speech

Johannes Müller, the founder of modern physiology and psychology, showed in 1848 that one of the biological mechanisms essential for human speech is the *supralaryngeal vocal tract*. The human vocal tract is essentially the top half of the airway that leads from the lungs into the atmosphere. The airway consists of several distinct segments. Extending from the lungs upward to about the level of the Adam's apple is the trachea or windpipe, capped by the larynx. The Adam's apple is itself the external sign of the thyroid cartilage of the larynx. Above the larynx the airway connects with the esophagus, which connects the stomach to the mouth. Ordinarily referred to as the throat, this region is the pharynx, where the food and air tubes come together. The pharynx branches into two passages, the oral and nasal cavities. As far as speech is concerned the important parts of the supralaryngeal vocal tract are the pharynx and oral and nasal cavities. The tongue, lips, the larynx (which can move upward or downward), and the velum, the soft flexible part of the palate that can close off the nose to the mouth, work together to change the shape of the supralaryngeal vocal tract.

Figure 2–1 shows the anatomy involved in the production of human speech. The lungs provide the energy that powers speech production. They also have a linguistic function; the flow of speech is segmented into sentence-length units by regulating the flow of air into and out of the lungs (Jones, 1932; Lieberman, 1967). A complete sentence is usually produced on one expiration. When we speak we take a volume of air into our lungs that is proportional to the length of the sentence we *intend* to utter (Lieberman and Lieberman, 1973).

The larynx is a device that can convert the steady flow of air out from the lungs into *phonation,* a periodic sequence of

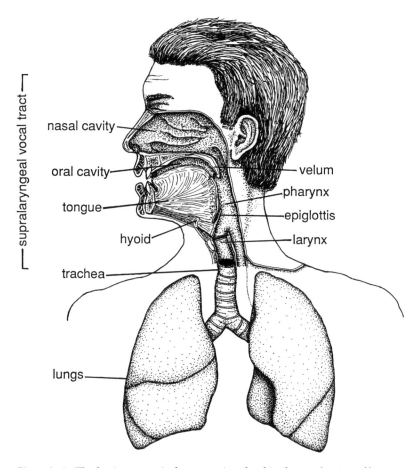

Figure 2–1. The basic anatomical systems involved in the production of human speech.

"puffs" of air. The puffs of air are produced as the larynx's vocal cords, which act as a complicated air valve, rapidly open and close. The puffs of air move up from the larynx into the vocal tract. Vowels are usually phonated, but they can also be produced with "turbulent" excitation when a person whispers. Consonants such as [v] and [m] are phonated. By convention, a letter in square brackets refers to a sound specified in the International Phonetic Alphabet (International Phonetic Association, 1949). The sounds of consonants generally correspond to their value in English orthography; for example [v]—*vat.* Vowels have a more complex relation to orthography; [i], for example, represents the vowel in the English word *see,* [u] represents the vowel of *boot,* [a] the vowel of *father,* [I] the vowel of *bit,* [e] the vowel of *met.*

The process by which speech is produced always involves a *source* of acoustic energy and a *filter* (Fant, 1960). The puffs of air that exit the larynx are similar to those produced by the reed of a woodwind instrument; they constitute a rich source of acoustic energy. The airflow from the laryngeal source enters the airway above it—the supralaryngeal vocal tract. The vocal tract acts as a filter in much the same way as the tube of a woodwind instrument. The musical quality of the notes produced by the instrument depends on the length and shape of the tube, which lets more acoustic energy through it at certain frequencies. The frequencies at which maximum acoustic energy will get through the vocal tract are called *formant frequencies.* Both woodwind instruments and the vocal tract act as filters, letting relatively more acoustic energy through at these formant frequencies.

The relationship that holds between the acoustic energy generated by the larynx, the supralaryngeal vocal tract, and the formant frequency pattern is similar to that which holds between sunlight, a stained-glass window, and the colors that one sees. The light produced by the sun is a *source* that has energy present at all the frequencies of the electromagnetic spectrum, ranging from low frequencies for red to high ones for blue. The stained-glass window is a *filter* that selectively blocks light energy at various frequencies. The particular colors that one sees correspond to the light energy at the frequencies that pass through the stained-glass filters; they are analogous to the speech sounds that correspond to various formant frequency patterns.

During the production of speech we continually change the shape and make small adjustments in the length of the vocal tract, thereby generating a changing formant frequency pattern. The formant frequency pattern is the primary determinant of the phonetic quality of the speech sounds that make up words. The vowel [e], for example, differs from [I] because it is made up of dissimilar formant frequencies. The high transmission rate of human speech is achieved through the generation of rapidly changing formant frequency patterns by the species-specific human vocal tract.

In contrast, the *fundamental frequency of phonation*, which is the rate at which the puffs of air occur, determines the "pitch" of a speaker's voice. The fundamental frequency pattern can convey linguistic information such as the pitch tones that differentiate words in languages such as Chinese, as well as the intonational breath-groups that segment the flow of speech into sentences and phrases.[1] The rapid change of formant frequencies, however, is the key to the speed of human speech.

Figure 2–2 illustrates the filtering effect of the supralaryngeal vocal tract. The plot at the top shows the spectrum of the energy produced by the laryngeal source. This is roughly similar to the raspy buzz you would hear if you held the reed of a woodwind instrument in your hand and blew through it. A spectrum is a visual representation of the distribution of energy with respect to frequency; a spectrum of the colors of the rainbow would show light energy present at the frequencies that correspond to the colors that we perceive, ranging from low frequencies for red to high ones for blue. The spectrum of the sound produced by the larynx shows that acoustic energy is present at the fundamental frequency of phonation, the lowest line in the graph at 500 hertz (Hz), and its harmonics. The plot therefore shows energy present at .5 kilohertz (kHz), 1.0 kHz, 1.5 kHz, 2.0 kHz, and so on (one kHz = 1,000 Hz). The height of the lines in the plot diminishes because the energy in the spectrum of the sound produced by the larynx generally falls off with frequency.

Speech Perception

The middle plot in Figure 2–2 shows the filter function of the supralaryngeal vocal tract for the vowel [i]. The first, second, and third formant frequencies, F1, F2, and F3, are the primary

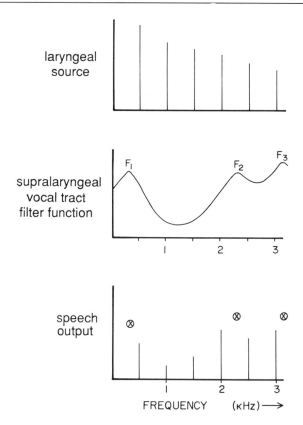

Figure 2–2. Graphs illustrating the source-filter theory of speech production; the horizontal axis for all three plots shows frequency in kilohertz (kHz). *Top:* The acoustic energy that would be present in the laryngeal source for a person phonating at a fundamental frequency of 500 Hz (0.5 kHz). The vertical lines show that energy is present *only* at the fundamental frequency of phonation—0.5 kHz—and its harmonics (frequencies that are integral multiples of F0) at 1 kHz, 1.5 kHz, 2.0 kHz, 2.5 kHz, etc. *Middle:* The vertical axis shows the relative amount of acoustic energy that would pass through the supralaryngeal vocal tract if it assumed the shape characteristic of the vowel [i] (the vowel of *meet*). The peaks in this filter function are the formant frequencies F1, F2, and F3. *Bottom:* The result of the supralaryngeal vocal tract filter's acting on the source. Energy is present only at the fundamental frequency and its harmonics. However, human listeners when presented with this signal "perceive" the supralaryngeal vocal tract's formant frequencies denoted by the circled Xs.

acoustic cues that make the vowel an [i]. They are the frequencies at which maximum acoustic energy will get through the supralaryngeal airway when the speaker places his or her vocal tract in the configuration that would produce the vowel [i]. (The filter function in Figure 2–2 corresponds to a male supralaryngeal vocal tract.)

The lower plot in Figure 2–2 shows the net effect of the supralaryngeal vocal tract filter on the laryngeal source. The frequencies of F1, F2, and F3 are marked by the circled Xs. The lower plot shows that there is no acoustic energy present at the formant frequencies. However, a human who heard the speech signal that corresponded to this graph would "hear" these formant frequencies. Almost 200 years of research demonstrate that human beings are equipped with neural devices that, in effect, calculate formant frequencies from the speech signal. We do this even when very little acoustic information is present, as is the case on a telephone. We appear to have innate knowledge of the filtering characteristics of the human supralaryngeal vocal tract—a complex neural formant frequency "detector" that calculates the formant frequencies on the basis of an internal representation of the physiology of speech production. Computer programs that go through a similar process are able to calculate the formant frequencies of unnasalized sounds with reasonable accuracy.

Nasalized Speech Sounds

One important point to note in connection with formant frequency detection is that it is very difficult for computer programs to calculate the formant frequencies of nasalized vowels. When the nose is connected to the rest of the supralaryngeal vocal tract, it introduces nasal formants and other effects that obscure the formant frequency patterns that differentiate vowels (zeros and increased bandwidth; see Lieberman and Blumstein, 1988). Human listeners have similar problems: nasalized vowels are misidentified more often than similar nonnasal vowels—30 to 50 percent more often (Bond, 1976). In other words, nasalized speech is inherently less perceptible.

Matched Brain Mechanisms for Decoding Speech

The speed of human speech derives from our having innate brain mechanisms that are adapted for speech perception. We unconsciously assign patterns of formant frequencies and other acoustic cues to discrete phonetic categories. One pattern of formant frequencies and phonation will be heard as a [b], another as a [p], another as an [a], and so forth. The brain mechanisms that do this appear to be matched, that is, tuned to respond to the particular acoustic signals that the human speech-producing anatomy can produce (Liberman et al., 1967; Liberman and Mattingly, 1985). Many other animals, including crickets, frogs, and monkeys, have brain mechanisms that are tuned to respond to their species-specific vocalizations. However, we appear to have a different, elaborated, and more efficient set of speech detectors. For example, on hearing 1/100 of a second of the start of a monosyllabic word such as *bee,* we "hear" the entire syllable. These speech detectors appear to be genetically transmitted; human infants behave in much the same way, shortly after birth, to these elementary aspects of speech. It is impossible even to survey the extensive research literature pertaining to the perception of human speech; more than one hundred independent studies are consistent with the premise that we are equipped with genetically transmitted neural devices that facilitate the perception of the particular sounds that occur in human speech (see Lieberman, 1984).

However, the perception of human speech also makes use of brain mechanisms that may have an auditory basis. For example, the phonetic distinction between the *stop consonants* [p] and [b] appears to depend on a basic constraint of the mammalian auditory system. Stop consonants such as [p] and [b] are produced at the start of a syllable by suddenly opening the lips. This results in a *burst,* a sudden sharp sound. The vocal cords of the larynx start phonation shortly after the burst for both [p] and [b]. Both sounds are *labial* stop consonants that have similar formant frequency patterns; the *voice-onset time (VOT),* the time interval between the burst and the start of phonation, is the primary acoustic cue that signals the [p] versus [b] distinction.

Labial stop consonants having a VOT between 0 and 25 milliseconds are heard as [b]s, those having longer intervals as [p]s. A
sharp "categorical" distinction occurs at about 25 milliseconds;
no intermediate sounds can be heard. The stops are all [b]s if
their VOT is less than 25 milliseconds, abruptly changing to
[p]s when their VOT exceeds 25 milliseconds. The experiments
of Patricia Kuhl, who has been working for years on the perception of speech by human infants, monkeys, and chinchillas as
well as human adults, show that they all appear to make use
of the same brain mechanism to keep track of this distinction—a sharp categorical distinction occurs at the minimum
time interval at which the auditory system can distinguish the
serial order of two dissimilar acoustic signals. For stop sounds
we have to be able to determine that the burst occurred before
the onset of phonation—the 25-millisecond categorical boundary corresponds to the minimum difference in time for which
we can make this decision. Similar effects differentiate the
sounds [t] from [d] and [k] from [g]. Human listeners can also
use other acoustic cues that derive from the special characteristics of the vocal tract to make these phonetic distinctions. The
brain mechanisms that appear to respond to these cues appear
to be special-purpose ones that exist only in human beings. The
process of speech perception again resembles other aspects of
human evolution—"newer" mechanisms have been added to
"old" ones to facilitate an important task (Lieberman, 1984).
Kuhl (1988) notes that general auditory constraints may have
determined the first phonetic—that is, linguistic—distinctions.
This is an eminently reasonable position, since all primates hear
and only one talks.

Vocal Tract Normalization

We humans perform some other remarkable feats as we listen
to speech. We have to estimate the probable length of a
speaker's supralaryngeal airway in order to assign a particular formant frequency pattern to a particular speech sound.
Different-length vocal tracts will have different formant frequencies; a short vocal tract will produce speech sounds that
have higher formant frequencies than a long vocal tract, just as

a piccolo and a bassoon produce musical notes with higher and lower pitches, respectively. The length of the supralaryngeal airway differs greatly in humans: in young children it is half as long as in adults. Adults' vocal tracts also vary in length, and because of this variation there is overlap between the formant frequency patterns that convey different speech sounds. If we interpreted the formant sounds of speech like the notes produced by woodwind instruments, the speech sounds of people with different vocal tract lengths would not have the same phonetic value. For example, the word *bit* spoken by a large adult male speaker can have the same formant frequency pattern as the word *bet* produced by a smaller male. Yet we "hear" the large persons's *bit* as *bit* rather than as *bet*.

Human listeners always normalize speech signals in terms of the probable length of a speaker's vocal tract. Experiments using artificial speech to confuse listeners show that listeners will interpret the same acoustic signal as a different vowel, depending on whether they believe the speech is being produced by a shorter or longer vocal tract (Ladefoged and Broadbent, 1957; Nearey, 1978). We perform this task unconsciously. The process is similar to the size normalization that occurs in vision, whereby we recognize someone's face independently of the size of the image projected on our retina. The vowel sound [i], which can be produced only by the human vocal tract, is an optimum cue for vocal tract normalization.

The imitative behavior of infants by age three months indicates that this process of vocal tract normalization is probably innate. Because it is impossible to ask infants what they hear, many studies of infant speech perception use the infant's imitation to infer perceptual categories. In one such study a three-month-old boy imitated the vowels that his mother produced (Lieberman, 1984, pp. 219–222). His supralaryngeal vocal tract was much shorter than hers, so it was physically impossible for him to produce the absolute formant frequencies of his mother's vowels. Acoustic analysis revealed that he produced frequency-scaled versions of his mother's vowels; his formant frequencies were proportional to the ratio between the length of his own supralaryngeal vocal tract and hers. This particular infant appears to have been somewhat precocious, but Patricia Kuhl's recent studies show similar behavior for many children after age four months. The children appear to normalize vowels in the same manner as adults (Nearey, 1978).

Automatization and Speech Production

Specialized brain mechanisms likewise seem to be involved in the production of human speech. Paul Broca in 1861 first noted that damage to certain parts of the brain disrupts the motor programs that control speech production. The articulatory maneuvers necessary to produce speech are among the most complex that humans perform. Until age ten, normal children cannot meet adult criteria for even basic maneuvers such as the lip positions that are necessary to produce different vowels (Watkins and Fromm, 1984). Humans must rapidly execute complex voluntary articulatory maneuvers involving the tongue, lips, velum, larynx, and lungs to produce a particular formant frequency pattern. These maneuvers are executed automatically; we don't have to think about what we are doing when we say *mama* or *disambiguation*.

The general process of *automatization* that allows us to learn to rapidly execute complex goal-directed patterns of motor activity is not specifically linguistic or even limited to humans. The motor activity that characterizes much of our daily life is "routine" and "automatic" because it is executed by reflexlike neural subroutines that execute goal-directed activity. The primary reflexes that are built into animals, such as the knee-jerk reflex—the innately determined movement of your lower leg that occurs when someone taps your knee in the appropriate place with a little hammer—elicit an automatic, rapid motor response. Human beings, like other animals, have many genetically transmitted primary reflexes that rapidly and automatically produce a motor response. However, these innate reflex-controlled motor responses do not include activities such as driving a car. Automatization (Evarts, 1973) converts a series of *learned* motor instructions into a "subroutine" that is stored in the motor cortex and executed as a complete whole.

An experienced car driver shifts gears using an automatized process—a complex sequence of muscular activity is executed without conscious thought. When you first learn to shift gears you must consciously execute each step; the process is slow, and you really can't think about other things. You are preoccupied with the mechanics of shifting. Once you learn to shift, the process becomes automatic—you don't consciously think about shifting, and the act is surprisingly faster. It is always disconcerting to "wake up" while you are driving and realize that you have no conscious memory of the past thirty minutes or so. The process of driving a car becomes so automatic that you can be driving along for miles without any conscious memory of what you have seen or done.

People have known about automatization for a long time. "Immediate action without conscious thought" was the goal of a seventeenth-century Japanese manual for learning to use the samurai sword. More recent though less entertaining studies have established the neural bases of automatization. Data derived from monitoring the electrical signals that control muscles and the neurons of the motor cortex show that reflexlike circuits are formed in the motor cortex when we automatize an action (for example, Evarts, 1973; Polit and Bizzi, 1978; Miles and Evarts, 1979). Edward Evarts was one of the first to show that monkeys "acquire" circuits in the motor cortex that facilitate particular goal-directed activities. He trained a monkey to grasp a movable handle firmly and keep it positioned on a visual target. External deflecting forces were then applied to the handle at unpredictable intervals. The monkey was rewarded when it kept the handle on target. The total time that elapsed between the perturbing force to the handle and the corrective muscle activity in a trained monkey's arm was about forty milliseconds, about twice as fast as the responses of untrained monkeys. This interval corresponds to the time it takes for an electrical signal to travel from the perturbed muscle to the motor cortex of brain, and for a correcting electrical signal from the motor cortex to travel back to the muscle.

Evarts (1973) also found neurons in the motor cortex of the trained monkey responding in this situation that were not present in untrained ones. Similar effects (except that motor cortex responses were not recorded) were found when the experiment was conducted with humans. The goal coded in an automatized motor cortex pathway may be a simple one—resisting the deflection of one's hand grasping a stick—or complex—shifting gears or producing the sound [b]. In all cases, however, the essential feature of automatization is that a complex motor routine, which involves the activity of many different muscles, is neurally coded and rapidly executed as an entity—a subroutine.

The automatized maneuvers that humans use in the production of speech are extremely complex. For example, the vowel [u] that occurs in the word *two* is produced with the lips rounded, that is, protruded and pursed. In contrast, the vowel [i] that occurs in *tea* is produced with the lips retracted. When adult speakers of English produce the word *two* they round their lips 100 milliseconds *before* they start to produce the vowel. When they produce *tea,* they don't round their lips at all. The speakers anticipate the occurrence of the vowel. This effect is general during the production of speech and is an example of *anticipatory coarticulation.* Adult speakers of Swedish, in contrast, coarticulate the [u] between 500 and 100 milliseconds ahead of its occurrence (Lubker and Gay, 1982). The difference is not the result of some genetic difference between speakers of English and Swedish. English-speaking three-year-olds, for example, fail to coarticulate their [u]s at all (Sereno and Lieberman, 1987; Sereno et al., 1987). Acquiring an English versus a Swedish accent involves learning these different automatized motor control patterns.

Additional data from experiments analogous to Evarts' provide some notion of the complexity of the automatized motor control patterns that underlie human speech. Vincent Gracco and James Abbs (1985) used a small electrical motor to apply a force to a speaker's lower lip to impede its closing. Speakers were asked to produce a series of syllables that started with [b], a *labial* stop consonant that is produced by closing the lips. The experimenters first occasionally applied a perturbing force forty milliseconds before the time when the lips would have otherwise reached a closed position. The speakers had no way of anticipating when this would be done. However, they compensated *within* the forty-millisecond interval between the perturbation and closing their lips by applying more muscle force to the lower lip, overcoming the motor's force. They also used a slight downward movement of the upper lip. The experimenters then shortened the time between the perturbing force and normal lip closure to twenty milliseconds. In this case, probably because there was insufficient time to overcome the perturbing force by means of lower lip action, the speakers compensated by a large downward deflection of the upper lip, extending the duration of the lip closure. The subjects thus used two different automatized motor response patterns that had a common goal—closing the lips to produce the consonantal stop closure. They appear to have a neural representation, probably involving Broca's area, of an abstract linguistic goal—closing one's lips for a [b].

Vocal Control in Chimpanzees

Broca's area, which is located near the cortical motor areas, seems to derive from premotor and prefrontal cortical areas that regulate orofacial muscular maneuvers in nonhuman primates. The premotor areas of the cortex are believed to govern the smoothness and sequencing of learned skilled movements (Luria, 1973). The prefrontal cortex is implicated in the regulation of organized conscious activity. However, although nonhuman primates have neocortical areas with a cytoarchitectonic structure similar to that of the human Broca's area, these parts of the nonhuman brain do not regulate their vocalizations. A functional Broca's area is present only in humans.

Nonhuman primates are unable to produce the voluntary muscular maneuvers that underlie human speech. For example, computer modeling and other studies show that although the chimpanzee airway is inherently unable to produce all the sounds of human speech, it could produce a subset of them—nasalized versions of vowel sounds such as [I], [e], [ae], and [U] and consonants such as [t], [d], [b], and [p] (Lieberman,

1968; Lieberman, Crelin, and Klatt, 1972; Richman, 1976). Therefore, on the basis of their anatomy it would appear to be feasible to train chimpanzees to produce a somewhat distorted version of most words as well as almost perfect though nasalized versions of words such as *food* or *bit*. However, over the past 300 years all attempts to teach chimpanzees to talk have failed. The last effort was mounted by Cathy and Keith Hayes, who raised a young chimpanzee named Viki with their infant son, Donald. Before training, Viki was "completely unable to make any sound at all on purpose" (Hayes and Hayes, 1951, p. 66). After intensive instruction, which involved manually opening and closing the chimpanzee's jaws in an attempt to get her to talk, she learned to make a sort of nasalized, breathy *ahhh* that indicated that she wanted various objects. She also produced rather indistinct versions of four words, including *mama* and *cup*. In contrast, Donald learned to imitate chimpanzee pant-hoots without any explicit instruction.

Furthermore, apes appear to have difficulty in the *intentional,* voluntary control of their vocal signals. Jane Goodall, for example, notes that chimpanzees are unable to suppress food-barks even when doing so is in their best interests. She explains:

> Chimpanzee vocalizations are closely tied to emotion. The production of a sound in the *absence* of the appropriate emotional state seems to be an almost impossible task for a chimpanzee . . . A chimpanzee can learn to *suppress* calls in situations when the production of sounds might, by drawing attention to the signaler, place him in an unpleasant or dangerous position, but even this is not easy. On one occasion when Figan [a chimpanzee at the Gombe Stream Reservation] was an adolescent, he waited in camp until the senior males had left and we were able to give him some bananas (he had none before). His excited food calls quickly brought the big males racing back and Figan lost his fruit. A few days later he waited behind again, and once more received his bananas. He made no loud sounds, but the calls could be heard deep in his throat, almost causing him to gag. (1986, p. 125)

As we noted in Chapter 1, the regulation of nonhuman primate vocalizations appears to derive from the cingulate cortex, basal ganglia, and brainstem rather than the neocortical areas that control speech in modern humans. Chimpanzee vocalizations seem to be tied to orofacial gestural patterns—the "grimaces" that Darwin noted in *The Expression of the Emotions in Man and Animals* (1872). The acoustic quality of a chimpanzee call, such as lowered formant frequencies resulting from lip rounding, thus derives from the orofacial expression. Attempts to train chimpanzees and monkeys in the laboratory to vocalize on command have succeeded in changing the rate at which or conditions in which they will produce stereotyped vocalizations (Pierce, 1985) but they have not been able to get them to produce any novel sounds.

The Evolution of the Human Supralaryngeal Vocal Tract

Nature is a miser; preadaptation is the evolutionary mechanism whereby old structures are remodeled at minimum expense to fulfill "new" tasks. The classic example that Darwin discussed in 1859 is the evolution of the respiratory system in air-breathing animals. A small modification to the swim bladders of fish was the starting point. The detailed anatomical studies by Sir Victor Negus between the two world wars show the series of small changes that produced the human larynx. The larynx generates the acoustic source that enters the supralaryngeal vocal tract. However, its primary life-supporting function in most animals is to protect the lungs from intruding foreign objects. It still has that function in humans: when we cough we are closing our larynx abruptly in a reflex action triggered by the intrusion of liquids or solid objects. Coughing reflects the most primitive function of the larynx; the larynges of the most primitive air-breathing animals, the lungfish, protected their lungs from water. Modern humans retain a larynx, but it can no longer protect our lungs from the intrusion of water when we swim. Too many changes have occurred because our larynx has become adapted for respiration and soundmaking. However, the starting point for the evolution of the human larynx was a small addition to some lungfish's larynx.

The opportunistic nature of evolution is apparent in the total "illogical" design of the human lungs. We have a system that works—we can breathe—but it is subject to all sorts of problems: collapse of the lungs, loss of elasticity that results in emphysema, and so on.

In the *Origin of Species*, Darwin commented on another example of illogical biological design: the peculiar anatomy of the human supralaryngeal vocal tract. He noted "the strange fact that every particle of food and drink which we swallow has to pass over the orifice of the trachea, with some risk of falling into the lungs" (1859, p. 191). One of the features that distinguishes humans from all other terrestrial mammals is that we are more liable to choke when we eat or drink. Solid objects or liquid can fall into the human larynx, obstructing the pathway to the lungs. Tens of thousands of people die each year because of the fact that our larynx is positioned low in our neck. You can feel your larynx when you swallow—the Adam's apple (the thyroid cartilage) marks its approximate position. In all other terrestrial animals it is positioned in the mouth, close to the bottom of the skull, where it can move upward through the mouth passage and forms a watertight seal with the entrance to the nose. The animal's larynx rises like a periscope; air goes through the raised larynx to the lungs, while food and water goes around it to the stomach. In correlation with the high position of the larynx, animals have long, thin tongues positioned entirely in their mouths. The animals' mouths, tongues, and laryngeal position are adapted to moving food and drink efficiently into their stomachs. Nonhuman mammals therefore can simultaneously breathe and drink.

Figure 2–3 shows the typical nonhuman airway. The long, relatively thin (in profile) tongue is positioned entirely within the oral cavity, where it forms the lower margin. The midsaggital view shows the airway as it would appear if the animal were sectioned on its midline from front to back. The long, relatively thin tongue positioned within the mouth matches the high position of the larynx. Contrast the nonhuman configuration with Figure 2–4, which shows the species-specific adult human airway. A round, "fat" tongue projects down into the throat. Half of the tongue forms the lower boundary of the mouth, half the

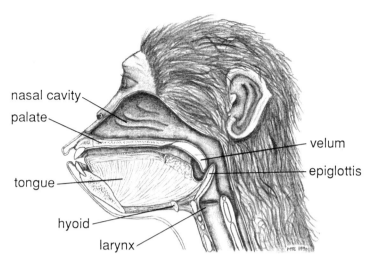

nasal cavity

palate

velum

epiglottis

tongue

hyoid

larynx

Figure 2–3. A typical nonhuman supralaryngeal airway: a chimpanzee. The tongue is positioned entirely within the oral cavity; the larynx is positioned high, close to the opening to the nose. The epiglottis and velum overlap to form a watertight seal when the larynx is raised, locking into the nose during feeding. The hyoid bone is connected to the larynx, jawbone, and skull by means of muscles and ligaments; it is part of the anatomical system that can raise the larynx.

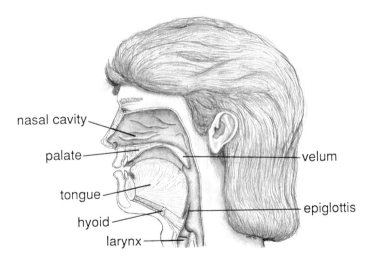

nasal cavity

palate

velum

tongue

hyoid

epiglottis

larynx

Figure 2–4. The supralaryngeal airway of an adult human being. The low position of the larynx makes it impossible for it to lock into the nose. The tongue has a very different shape from those of all other terrestrial mammals; its posterior contour is almost round in this lateral view and forms both the floor of the oral cavity and the front part of the pharynx.

anterior boundary of the pharynx. The human larynx cannot reach the opening to the nose because it is positioned at the lower end of the tongue. Air, liquids, and solid food make use of the common pharyngeal pathway, sliding past the laryngeal opening to the lungs. Humans thus are more liable than other terrestrial mammals to choke when they eat because food can fall into the larynx, obstructing the pathway into the lungs. The human configuration is also less efficient for chewing because the palate (the roof of the mouth) and mandible (lower jaw) are relatively shorter than in nonhuman primates and archaic hominids. Chewing is faster with longer mandibles that accommodate larger, stronger muscles and larger teeth; larger tooth areas also yield more efficient chewing (Manley and Braley, 1950; Manley and Shiere, 1950).

The reduced length of the modern human palate and mandible also crowds our teeth, presenting the possibility of infection from impacted wisdom teeth—a condition that was usually fatal until the introduction of anesthesia in dentistry in the nineteenth century. Impacted wisdom teeth also result in infections that were difficult to treat until antibiotics became available in the 1940s. How many people do you know who have had their wisdom teeth removed? They probably would have died without modern dental care. The only thing to which the adult supralaryngeal vocal tract is better suited in humans than in other animals is the production of the sounds of human speech. The problem that Darwin noted thus follows from preadaptation; a system that was initially adapted for breathing and eating was preadapted for a new function—speech—by changes in the shape of the tongue, the position of the larynx, and the supporting skeletal structures. Again, the system works, but it is not optimal.

Human newborn infants have a supralaryngeal vocal tract similar to that in nonhuman primates (Negus, 1949; Crelin, 1969; Lieberman and Crelin, 1971). Retaining the nonhuman supralaryngeal airway during early infancy contributes to biological fitness because newborn infants would not be able to talk even if their vocal tract were fully developed. It is therefore more advantageous to retain the nonhuman airway, which reduces the risk of choking to death on liquids. As infants grow, their palates move backward in relation to the base of the skull. The base of the human adult skull is restructured in a manner unlike that of all other mammals to achieve the adult human supralaryngeal airway (Laitman and Crelin, 1976).

The Selective Advantages of Human Speech

Vocal communication clearly exists in all living primates. Furthermore, Negus (1949) demonstrated that the larynges of social mammals are adapted for phonation at the expense of respiration. Hence a stage in hominid evolution in which communication was entirely gestural is most unlikely. Human speech contributes to our biological fitness in at least three ways:

1. *Nonnasal sounds.* The minimum contribution to fitness of the human supralaryngeal airway is its ability to produce sounds that are *not* nasalized. The velum in the human airway closes off the nasal cavity from the rest of the airway. The sharp bend in the human supralaryngeal vocal tract and the short distance spanned by the velum make it possible to seal the nose off. Nasalized sounds occur when we do not seal off the nose; as was noted earlier, it is harder to determine nasalized formant frequency patterns. They are misidentified by human listeners 30 to 50 percent more often than are nonnasalized sounds (Bond, 1976). The higher nasal error rate obviously interferes with the effectiveness of vocal communication, and human languages tend to avoid using nasal sounds (J. Greenberg, 1963).

2. *Quantal sounds.* There are further phonetic advantages that derive from the morphology of the modern human supralaryngeal airway. The round human tongue moving in the right-angle space defined by the palate and spinal column can generate formant frequency patterns that define quantal sounds (Stevens, 1972): the vowels [i], [u], and [a] (the vowels of the words *meet, boo,* and *mama*) and consonants such as [k] and [g]. Quantal sounds facilitate vocal communication in two ways.

First, quantal sounds provide *acoustic salience;* that is, their formant frequency patterns yield prominent spectral peaks (formed by the convergence of two formant frequencies; Fant,

1956) that make it easier to perceive these sounds, just as, in the domain of color vision, saturated colors are easier to differentiate than muted ones. Acoustic salience derives from the ability to produce an abrupt change in the cross-sectional area of the supralaryngeal vocal tract. In all mammals' vocal tracts it is possible to do this by moving the tongue blade (tip) up against the palate to produce dental consonants such as [d], [s], and [t]. In all primate vocal tracts the lips can close to produce labial consonants such as [b], [p], and [m]. The right-angle bend in the vocal tract of modern humans and some archaic fossil hominids permits sounds such as the velar consonants [g] and [k] and vowels such as [a], [u], and [i], which also have prominent spectral peaks. These sounds—the vowels [a], [u], and [i] and the labial, dental, and velar consonants—are better suited for vocal communication than are other sounds. They occur more often in different human languages (J. Greenberg, 1963).[2] Moreover, children (Olmsted, 1971) and adults (Peterson and Barney, 1952; Miller and Nicely, 1955) make fewer errors in identifying these sounds. The error rate for misidentification of the vowel [i] is particularly low—for adults, 6 errors in 10,000 trials (Peterson and Barney, 1952).

The vowel [i] also is an optimum cue for the process of vocal tract normalization, which is absolutely necessary for the formant frequency encoding that gives human speech its high data transmission rate. It is possible to use other sounds to guess at the length of a speaker's vocal tract, but [i] is the best vowel sound (Nearey, 1978). In other words, having a vocal tract that can produce the vowel [i] facilitates the process that gives human speech its high data transmission rate.

Second, quantal sounds enhance vocal communication through their *acoustic stability.* Kenneth Stevens (1972) has demonstrated that the human vocal tract allows us to generate the prominent spectral peaks of quantal sounds without having to be extraordinarily precise when we position our tongue; we can be somewhat sloppy and still produce signals that are acoustically distinct. Stevens used mechanical and computer models of the human vocal tract and systematically moved the tongue body (the main part of the tongue). Because the values of formant frequencies depend on the length and shape of the vocal tract (Fant, 1960), models of the vocal tract can be used to calculate formant frequencies. Stevens noted that the round human tongue moves as an almost undeformed body when we produce different vowels and velar consonants such as [g] and [k]. Stevens found that small errors in positioning the tongue to produce the vowels [i], [u], and [a] and velar consonants such as [g] did not radically change their formant frequencies, because changes in the shape of the oral cavity would be offset by corresponding changes in the shape of the pharynx— if the tongue body moved forward, shortening the length of the oral cavity automatically produced a somewhat longer pharyngeal cavity. Stevens also modeled the quantal sounds that can be produced by all primate supralaryngeal vocal tracts, consonants such as [b], [p], [d], [t], and [s]. These sounds also are acoustically salient and stable, because they are generated by the lips and tongue blade (the tongue tip), which are functionally similar in nonhuman primates and humans.

3. *Speech encoding.* A data transmission rate three to ten times faster than that of any other primate is obviously a strong selective advantage. The value of rapid communication probably increases with cultural complexity. However, rapid speech communication would contribute to biological fitness even in a "simple" cultural setting. Consider the selective value of being able to communicate the encoded message "There are two lions, one behind the rock and the other in the ravine" in the same time as the unencoded "lliiooonn rooockkk."

One can argue that other paths for rapid vocal communication could have evolved, such as the rapid transitions in pure tones that occur in communication among birds (Greenewalt, 1968). However, as primates we are subject to the constraint that we cannot identify nonspeech sounds at rates that exceed seven to nine items per second (Miller, 1956). Speech encoding by means of formant frequency transitions and other related cues (such as integrated spectral onset cues; Blumstein and Stevens, 1979) is the basis for rapid vocal communication in anatomically modern humans, and this process is closely tied to the properties of the human supralaryngeal vocal tract.

Is Speech a Selective Force for the Evolution of the Vocal Tract?
Although there is general agreement among scholars that the human vocal tract is necessary to produce the full range of human speech sounds, many have suggested that the singular anatomy of the human vocal tract and skull base has absolutely nothing to do with speech production. According to this view, the morphology of the skull base and vocal tract derives from adaptations to upright posture or the increase in the size of the hominid brain. Studies of developmental anomalies of the vocal tract constitute experiments in nature that allow us to refute these claims.

As we noted above, human newborns have an airway that is very similar to a nonhuman primate's. During normal development the palate gradually moves backward along the bottom of the skull. Major changes occur by age three months, but the process continues at a rapid pace until about age five and does not really end until adolescence. In people with Apert's and Cruzon's syndromes the supralaryngeal vocal tract is anomalous because the palate continues to move back along the sphenoid bone on the base of the skull, past its normal position. Karen Landahl and Herbert Gould (1986) use acoustic analysis, psychoacoustic tests, and computer modeling based on cephalometric x rays (which are carefully calibrated to allow precise measurements) to study the effect of the supralaryngeal vocal tract on speech. These show that the phonetic output of subjects with these syndromes is limited by their supralaryngeal vocal tracts. Apert's and Cruzon's subjects attempt to produce normal vowels but are unable to produce the normal range of formant frequency values. Psychoacoustic tests of their speech production yield a 30 percent error rate for identification of their vowels. Acoustic analysis shows that they are unable to produce normal [i] and [u] vowel sounds.

Figure 2–5 shows the results of a computer modeling of the supralaryngeal vocal tract of an Apert's subject. The computer has calculated the formant frequencies of the vowels that the speaker *could* have produced if he had full muscular control of his vocal tract. The technique therefore allows us to assess the limits that the speaker's anatomy places on speech production,

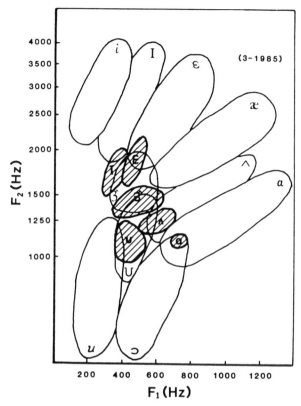

Figure 2–5. The first two formant frequencies of a vowel, F1 and F2, determine its phonetic quality. The open loops show the combinations of F1 and F2 (plotted on the two axes of the graph) that correspond to the vowels of English for normal adult speakers (Peterson and Barney, 1952). The shaded loops show the formant frequency combinations that the anomalous supralaryngeal vocal tract of a patient having Apert's syndrome can produce. The anomalous vocal tract inherently cannot produce any [i] vowels. (After Landahl and Gould, 1986)

independent of any possible motor control deficits. The loops of vowels produced by the normal subjects in Gordon Peterson and Harold Barney's (1952) definitive study are shown on the F1 versus F2 formant frequency plot. The values of F1 are plotted on the horizontal, F2 on the vertical axis. The first two formant frequencies of a vowel essentially determine its phonetic quality. Any vowel having first and second formant frequencies, F1 and F2, in the [I] loop would, for example, produce an [I] vowel (the vowel of *bit*) for some speaker. These loops, therefore, indicate the formant frequency combinations that will produce the vowels of English. The Peterson and Barney "loops" do more than this; they have been replicated many times for different languages and appear to show the range of vowels that can occur for all human languages. Different human languages divide the "vowel space" up in different ways, but the total space is about the same for all languages that have been analyzed (Lieberman and Blumstein, 1988).

The shaded areas toward the center of these loops represent the formant frequency patterns that the Apert's supralaryngeal vocal tract can produce when the experimenters attempted to perturb it toward the best approximation of a normal vocal tract's configuration for each vowel. Note that the modeled Apert's vocal tract cannot generate any formant frequencies that fall into the [i] loop and barely grazes the [u] loop. The possible [u] vowel would correspond to that of a person having a very short vocal tract and is inconsistent with the possible [I], [ɛ], [Λ], and [a] vowels, which are ones that would normally be produced by persons having very long vocal tracts. Therefore, we can conclude that the Apert's vocal tract cannot produce normal [i]s or [u]s. Because the computer-modeled plot in which the Apert's vocal tract was pushed to its phonetic limits is virtually identical with the formant frequencies that the Apert's subject *actually* produced, we know that he was using his vocal tract to its maximum phonetic capacity.

These data and data from other pathologies and from normal infants and apes are consistent with the hypothesis that natural selection for the ability to produce the vowel sounds [i] and [u] played a part in the evolution of the human supralaryngeal vocal tract. They show that in its normal configuration the human supralaryngeal vocal tract yields the maximum formant frequency range. Configurations in which the palate is anterior (Lieberman et al., 1972) or posterior (Landahl and Gould, 1986) to its "normal" adult position yield a reduced range of formant frequency patterns. Other factors, such as the redistribution of the forces involved in chewing, may have been involved in the restructuring of the human palate and mandible. However, it is unlikely that upright bipedal locomotion or brain size were major factors, since the anomalous Apert's and Cruzon's positions furthermore do not appear to be detrimental to upright bipedal posture and locomotion, nor is brain size reduced.

Classic Neanderthals

Ultimately the fossil record that pertains to human evolution has to be interpreted if we want to make any definite statements about the evolution of human speech. In the past twenty years a great deal of effort and debate has focused on reconstructions of the vocal tracts of various fossil hominids. Curiously, the debate has been most intense concerning the Neanderthal fossils, which were the first to be recognized as archaic hominids. They are also the largest group of recent hominids who clearly differ from anatomically modern *Homo sapiens.* They lived throughout Europe and Southwest Asia (what we now call the Middle East) until about 35,000 years ago, when they disappeared.

In the 1960s and 1970s some studies claimed that the anatomy of Neanderthal hominids fell into the range of variation of modern human beings (see Day, 1986). However, these studies applied the term Neanderthal to *all* fossil hominids who lived between about 125,000 and 35,000 years ago. Some of these fossils are quite similar to the skeletal remains of modern humans, but the true Neanderthals differ in fundamental ways. Eric Trinkhaus and William Howells (1979) have noted that "the Neanderthal complex of traits is simply not there" in even the earliest examples of modern *Homo sapiens* (p. 129). Many skeletal features of Neanderthal fossils such as finger bones, ankles, shoulder blades, and legs differ radically from those of modern humans. Neanderthals also had a nonhuman supralaryngeal vocal tract, although they walked perfectly upright and had brains that were as large as or larger than ours (Holloway, 1985)—a fact that eliminates these two factors as possible reasons for the shape of the modern human airway.

The first attempt to reconstruct fossil human supralaryngeal vocal tracts must be credited to Victor Negus and Arthur Keith.

In *The Comparative Anatomy and Physiology of the Larynx* (1949), Negus first demonstrated that the supralaryngeal vocal tracts of all other mammals differ from those of adult humans. Negus and Keith then reconstructed the supralaryngeal airways of a Neanderthal fossil. They concluded that the Neanderthal lacked a human tongue and pharynx; the position of its larynx and the shape of its tongue were instead closer to those of a chimpanzee. The details of their Neanderthal reconstruction are unfortunately not specified. However, it is significant that Negus and Keith, who were unaware of the role of the round human tongue and pharynx in human speech, reached the same conclusion as Edmund Crelin and I did in 1971. Crelin's reconstruction, based on comparative studies of the basicranium and mandible of human newborns, chimpanzees, and the La Chapelle-aux-Saints Neanderthal fossil, yielded the nonhuman configuration that is pictured in Figure 2-6.

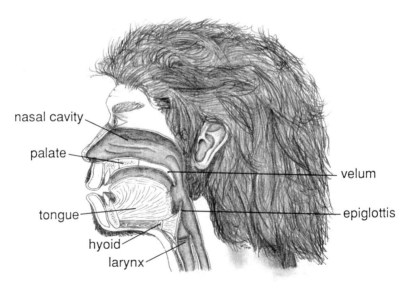

Figure 2–6. The reconstructed airway of the La Chapelle-aux-Saints Neanderthal fossil. Note the shape of the tongue, which is almost entirely within the mouth. The larynx is positioned high, close to the entrance to the nasal cavity. The supralaryngeal vocal tract is closer in form to a nonhuman primate's than to that of a normal modern human being.

The computer-modeling technique noted earlier was used to determine the phonetic output of the reconstructed Neanderthal airway. Computer modeling is not strictly necessary; since the vocal tract is similar to a woodwind instrument, the phonetic limitations of the reconstructed Neanderthal airway could have been assessed by making tubes that had the appropriate shapes and exciting them with a reed. However, computer modeling is quicker and more accurate. The output of the Neanderthal airway is quite similar to that of nonhuman primates and human newborns (Lieberman, 1968; Lieberman, Crelin, and Klatt, 1972). Figure 2-7 shows the output of the computer modeling. The loops again show the formant frequency combinations of F1 and F2 that will produce the vowels of English. The modeled Neanderthal vowels are indicated by the circled numbers. They were achieved by perturbing the reconstructed Neanderthal vocal tract into configurations that would best approximate those necessary to produce these sounds (Lieberman and Crelin, 1971). The Neanderthal vocal tract could not form the configurations that are necessary to produce [i], [u], and [a] vowels. Neanderthal speech is also nasalized and therefore would be subject to higher phonetic errors. If Neanderthal hominids had had the full perceptual ability of modern human beings, their speech communications, at minimum, would have had an error rate 30 percent higher than ours.

The computer modeling does not show that Neanderthal hominids totally lacked speech or language; they had the anatomical prerequisites for producing nasalized versions of all the sounds of human speech save [i], [u], [a], and velar consonants and probably had fairly well-developed language and culture (Crelin and Lieberman, 1971). The La Chapelle-aux-Saints fossil was described in detail by Marcellin Boule (1911–1913), who was the leading French anthropologist of the period; it is the "type" Neanderthal fossil. The La Chapelle Neanderthal man was buried about 50,000 years ago in a small cave in southwestern France that contained many stone tools and animal bones. By the time of his death he had lost most of his teeth and had arthritis. It is probable that Neanderthal hominids had a fairly advanced culture in which elderly, infirm individuals could sur-

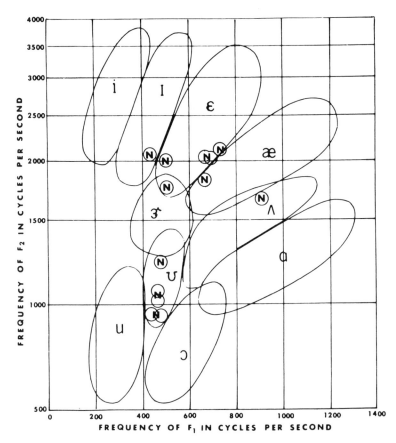

Figure 2–7. The computer-modeled formant frequencies of the vowels that could be produced by the reconstructed supralaryngeal vocal tract of the La Chapelle-aux-Saints Neanderthal fossil. The circled Ns denote the modeled vowels. The open loops containing phonetic symbols show the total range of vowel formant frequencies of English produced by normal adult speakers. Note that the Neanderthal vocal tract inherently cannot produce the vowels [i], [u], and [a]. The modeled Neanderthal was given a modern human larynx and hyoid; the phonetic limits follow from his nonhuman tongue. (After Lieberman and Crelin, 1971)

vive. In this sense they differed dramatically from present-day chimpanzees, who do not aid infirm members of their group (Goodall, 1986). Complex stone tools and the use of fire were other features of Neanderthal life. Because they lived in a cold climate, they probably wore clothes. Some scholars believe that they made fairly complex wood-framed skin shelters and produced art (Marshack, 1990). However, though Neanderthal hominids are not the savage brutes depicted in movies, they did not have human speech, because they did not have a human vocal tract.

Giving a Modern Vocal Tract to a Neanderthal

Among many detailed discussions of the skeletal features involved in reconstructing the supralaryngeal airway (see Lieberman, 1984, for references), some studies (Falk, 1975; Dubrul, 1977; Arensburg et al., 1989) have claimed that Neanderthal hominids had a completely modern human supralaryngeal airway that could produce the full range of human speech sounds. It is easy to demonstrate that this is impossible and in doing so to point out the key factors that differentiate the human supralaryngeal vocal tract and basicranium from those of archaic hominids.

If we instead claim that Neanderthals had human speech, we have to give the Neanderthal fossil a modern human supralaryngeal vocal tract—one with a curved tongue body that forms both the floor of the oral cavity and the anterior wall of the pharynx. Many independent studies (for example, Russell, 1928; Perkell, 1969; Ladefoged et al., 1972; Nearey, 1978) have demonstrated that the posterior contour of the human tongue is almost circular. As a result, the span of the tongue within the mouth is equal to the vertical distance between the roof of the mouth (the hard palate) and the top of the larynx (the epiglottis). This is evident in Figure 2–4, which shows a modern human supralaryngeal vocal tract.

The larynx in a modern human consequently is positioned low, but it is positioned within the neck; the laryngeal opening at rest is between the fifth and sixth cervical vertebrae. Figure 2–8 shows the La Chapelle-aux-Saints skull with a modern human vertebral column and tongue. The modern human tongue must span the long Neanderthal oral cavity. Since the distance from the teeth to the back of the mouth is great, the radius of the human tongue shape that is fitted to the Neanderthal skull must also be great; otherwise, the Neanderthal would be unable

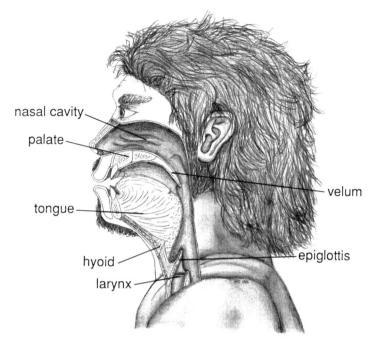

Figure 2–8. The La Chapelle-aux-Saints Neanderthal fossil with a modern human vocal tract that would have allowed him to produce the full range of human speech sounds. This entails having a tongue with a round posterior contour, half being in the mouth and half in the pharynx. The great length of the classic Neanderthal mouth results in an equally long pharynx; as a result, his larynx would be positioned in his chest.

to swallow food. A long oral cavity requires an equally long pharynx because the human tongue's posterior contour is round. The result is a larynx positioned below the neck in the Neanderthal chest. The reconstruction yields an impossible creature: no mammal has its larynx in its chest. Hence we must conclude that the Neanderthal fossil could not have had a modern human tongue or vocal tract.

This reconstruction of a human-Neanderthal supralaryngeal vocal tract reflects an attempt to accommodate a human tongue to the long Neanderthal mouth in every possible way: the tongue and pharynx are those of an adult female (there is a slight degree of dimorphism in modern humans with regard to the tongue shape, and males tend to have a slightly longer

pharynx length; Fant, 1960); and the Neanderthal vertebral column was shorter than that of modern humans (Boule and Vallois, 1957). All these factors would have placed the larynx still
lower in the Neanderthal chest if we insist, as was once proposed, that a classic Neanderthal hominid is no different from
other passengers who may be riding the New York subway.

A Neanderthal could not have had a modern human vocal
tract, because a long nonhuman mouth precludes having a human tongue. Victor Negus (1949, p. 26) recognized this fact and
ascribed the descent of the human larynx to the "recession of
the jaws" (the movement of the palate and mandible back along
the base of the skull). However, some anthropologists continue
to attempt to reconstruct fossil hominids without taking these
anatomical constraints into account. A research team headed
by Baruch Arensburg (1989) has claimed that the Kebara Neanderthal fossil, which has a long mandible, had a completely
modern vocal tract. They base this claim on the fossil's hyoid
bone, which they claim is similar to that of a modern human
being. But the hyoid bone is not rigidly attached to any other
bone of the body or to the larynx; therefore, it is difficult to
see how it alone could indicate the shape of the vocal tract
(Lieberman et al., 1989). Moreover, it is not clear that the Kebara hyoid is actually similar to a modern human's hyoid.
Arensburg and his colleagues base their claim on statistical
comparisons, but their own analysis places the Kebara hyoid
outside the human range (2 standard deviations away from
human hyoids on three of the five measures that they note).
Furthermore, the significance of these metrics is doubtful, since
the hyoids of living pigs are closer to those of humans if we
use the same criteria (Laitman et al., 1990). Thus we could as-
sert with equal confidence that pigs can talk.

Quantitative Methods for Reconstructing Vocal Tracts

Although the qualitative procedure discussed above suffices to
show that the Neanderthal skull could not support a modern
supralaryngeal vocal tract, it is insufficient to elucidate the full
range of variation that can be found in the fossil record. Fortunately, quantitative methods have been developed that allow

us to make reasonable inferences about the speech capabilities of many hominid fossils. These methods use procedures that are common in the reconstruction of the soft tissue of extinct animals. Relationships between skeletal structures and soft tissue are first established for living animals that resemble the extinct species. The skeletal structures of fossils are then compared with those of the living creatures, allowing inferences to be made about the soft tissue that the fossil probably had.

The method used for the reconstruction of vocal tracts thus is not very different from that used to reconstruct the way that dinosaurs walked. Muscles always leave marks on bones, and living species can be found that resemble the extinct fossils. The relationships between the leg bones and muscles of lizards serve to guide the reconstruction of dinosaur leg muscles. In this case the relationship between the skull and vocal tract in present-day apes and human adults and newborns provides the basis for correlating skeletal structures with soft tissue. Figure 2–9 shows the skeletal features that have been used in these studies; the relevant measurements are shown on a chimpanzee skull (Laitman, Heimbuch, and Crelin, 1979; Laitman and Heimbuch, 1982). Statistical analyses show that the distance between the hard palate of the mouth and the vertebral column (the distance between points B and E) and the flexure of the skull base correlate with the position of the larynx and the shape of the tongue. The flexure of the skull base (the basicranium) is defined by the angle formed between the base of the occipital bone and the entrance to the nasal cavity (the angle between lines BC and CE).

The distance between the end of the palate and the spinal column has to be long to accommodate the larynx in the nonhuman primate and human newborn vocal tract when it is raised into breathing position. The shallow angle between lines BC and CE likewise corresponds to the orientation of the pharyngeal muscles in the nonhuman vocal tract. In contrast, in adultlike modern humans (from about age ten on) the distance between points B and E is short, and the angle between lines BC and CE is acute. It is impossible to give a fossil skull that has these modern human features a nonhuman vocal tract; there

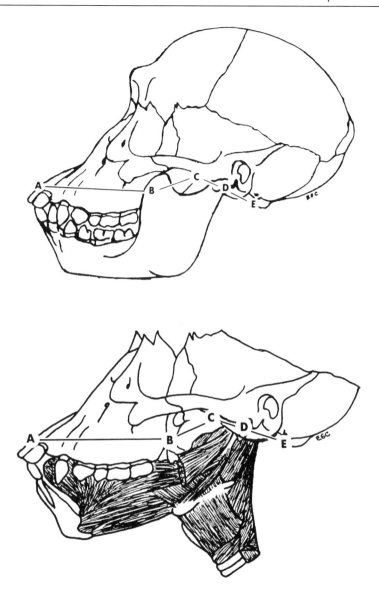

Figure 2–9. Side views of an adult chimpanzee skull with and without muscles of pharynx and tongue, showing craniometric points appropriate for quantitative comparisons of the relationship between the base of the skull and supralaryngeal vocal tract: A—prosthion; B—staphylion; C—hormion; D—sphenobasion; E—endobasion. (After Laitman, Heimbuch, and Crelin, 1978)

simply is not enough space to position the larynx close to the base with the pharynx behind it.

The Origins of Modern Humans and Modern Speech

It is apparent that the anatomical specializations that confer human speech differentiate modern humans from archaic hominids (Lieberman, 1984, 1989). All the fossil remains of early anatomically modern *Homo sapiens* that have so far been found have skulls that would have supported a modern supralaryngeal vocal tract. The early course of human evolution clearly involved anatomical adaptations that enhanced aspects of human behavior such as chewing, upright posture, bipedal locomotion, and precise hand maneuvers, but the last phase of hominid evolution appears to involve among other things the appearance of human speech. We can therefore trace the evolution of modern human beings in the fossil record through the features of the skull that yield human speech. These features also serve as an index for the presence of the brain mechanisms which are also necessary for speech, and which, as the following chapters argue, are key elements in the evolution of human language, cognition, and culture.

The general evolutionary order of fossil hominids, shown in Figure 2–10, starts with *Australopithecus afarensis* and other australopithecines who resemble living apes in many ways except for having somewhat larger brains, more modern hands, and pelvic regions, legs, and feet adapted for upright bipedal locomotion. However, they are not as exclusively adapted for upright bipedal locomotion as modern humans (Stern and Susman, 1983). Australopithecines may have spent a good part of their lives climbing trees. The skull bases of australopithecines are almost identical with those of present-day apes, indicating that they had apelike vocal tracts. In this connection, the vocal anatomy of living nonhuman primates would be sufficient for complex vocal communication *if* they could achieve voluntary control of the complex motor controls necessary for human speech. Modulations of pitch and some formant transitions and patterns can and do seem to occur in chimpanzee calls (Goodall,

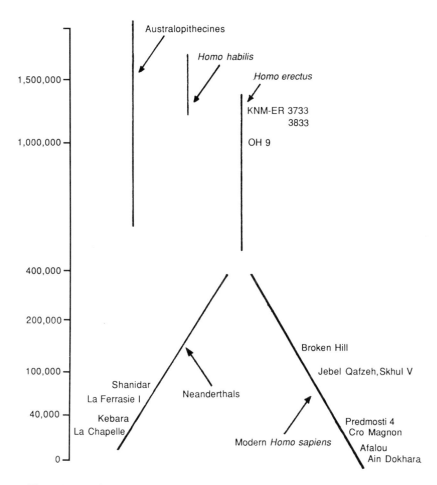

Figure 2–10. The evolutionary order of hominid fossils, showing the comparatively recent origin of anatomically modern *Homo sapiens.* Jebel Qafzeh and
Skhul V are the earliest accurately dated anatomically modern hominids; they
had modern supralaryngeal vocal tracts. Speech production would have been
facilitated at the expense of vegetative functions such as chewing and swallowing. They must have had brains that would allow the voluntary production of
the complex articulatory maneuvers that are necessary to produce human
speech. Neanderthal hominids form a side branch.

1986). Thus it is clear that the evolution of the brain mechanisms that facilitate *voluntary* control of vocal communication is one of the keys to the evolution of human speech. Australopithecines may resemble present-day chimpanzees in this respect; they may not have been able to produce vocalizations that were decontextualized from gestural displays. Therefore, gestures may have been the primary mode for australopithecine referential communication (Hewes, 1973). However, we can only speculate as to whether they had brains that would allow voluntary control of speech.[3]

The first major change from the nonhuman vocal tract that characterizes all other living terrestrial mammals (Negus, 1949) occurs in *Homo erectus* (Laitman and Reidenburg, 1988). The fossils that typify *Homo erectus* have larger brains than australopithecines. The flexure of the basicranium in *erectus* fossils such as KNM-ER-3733 is also greater than that in living apes or australopithecines, indicating that they had a larynx positioned lower in the neck. An experiment that directly addresses the relationship between the position of the larynx and the flexure of the basicranium shows that they are tightly coupled. When the angulation of the skull base of an infant mouse is increased by means of microsurgery, the larynx falls to a lower position (Laitman and Reidenberg, 1988). The hypothetical *Homo erectus* supralaryngeal airway would not suffice to produce the useful quantal speech sounds that humans can make, but neither would it impede the swallowing of food as much as the modern human configuration does. The long mandible and corresponding facial architecture would be well suited for chewing. The lower larynx position probably evolved to facilitate mouth breathing, which is an advantage for aerobic activities.

The brain mechanisms that regulate respiration would have had to allow voluntary mouth breathing to take advantage of the change in the position of the larynx. Table 2–1 summarizes these possibilities. Various studies, discussed in the next chapter (Kimura, 1979; Lieberman, 1984; MacNeilage, 1987), have proposed that the brain mechanisms that regulate the production of human speech may derive from ones that originally evolved to facilitate skilled one-handed work—primarily with

Table 2–1. Skeletal and neural adaptations for speech

Australopithecines

Apelike vocal tract
No vegetative deficits in comparison to nonhuman primates

Homo erectus

KNM-ER 3733
Slight lowering of larynx
Voluntary mouth breathing
Speech limited to range of sounds of nonhuman primates
No deficits in swallowing or chewing in comparison to nonhuman
  primates
Brain mechanisms to allow voluntary mouth breathing
Brain mechanisms for automatized speech control?

Anatomically modern *Homo sapiens*

Broken Hill
Long palate but flexed skull base
Vocal tract generating all sounds of human speech, but with less stability
Deficits in swallowing—choking
Brain mechanisms to allow voluntary mouth breathing
Brain mechanisms for automatized speech control

Jebel Qafzeh and Skhul V
Modern skull base
Modern vocal tract, all sounds of human speech
Deficits in swallowing—choking
Deficits from infected impacted teeth
Brain mechanisms to allow voluntary mouth breathing
Brain mechanisms for automatized speech control

Neanderthal

La Chapelle-aux-Saints
Long palate, shallow basicranial angle
Nasal speech, no quantal vowels
No deficits in swallowing
No deficits from infected impacted teeth
Better chewing
Brain mechanisms for automatized speech control?

tools. Since *Homo erectus* hominids used and made tools, they may have possessed at least rudimentary voluntary speech motor ability.

The African Broken Hill fossil, which is considered to be an example of archaic *Homo sapiens* (Day, 1986), has a modern cranial base angle. However, the palate is longer than that of modern humans, so chewing would have been more efficient than in modern humans. On the other hand, it would have been susceptible to greater risk of death from swallowing than would early *Homo erectus* fossils such as KNM-ER-3733, australopithecines, and apes. Its supralaryngeal vocal tract would have been able to produce the full set of quantal speech sounds, as well as unnasalized speech, though with less stability than a fully modern vocal tract. There would have been a relatively lower biological fitness *unless* Broken Hill could have used its vocal tract to produce speech. We can therefore conclude that Broken Hill also had a brain that allowed voluntary control of speech. The brain mechanisms that allow automatized control of speech production probably were present to some degree.

A completely modern supralaryngeal vocal tract is present about 100,000 years ago in the Jebel Qafzeh VI and Skhul V fossils from Israel. The length of the palate is similar to that of present-day humans, and the vocal tract would have produced quantal speech sounds that were stable. Recent theories propose that anatomically modern *Homo sapiens* originated in Africa somewhere between 100,000 and 400,000 years ago, subsequently dispersing through the Middle East to Europe and Asia (Stringer and Andrews, 1988). The presence of a functionally modern vocal tract in the African Broken Hill fossil 125,000 years ago and its retention and elaboration in Jebel Qafzeh VI and Skhul V 100,000 years ago are consistent with this theory.[4]

Conversely, the extinction of Neanderthal hominids may derive from their having lacked human speech. At minimum they would have had less efficient vocal communication—more confusable speech, and perhaps a very slow rate. Any of these deficits would suffice to explain their replacement by our ancestors. As Darwin put it, in the "struggle for life, any variation, however slight and from whatever cause proceeding, if it be in any degree profitable to an individual of any species, in its infinitely complex relations to other organic beings and to external nature, will tend to the preservation of that individual, and will generally be inherited by its offspring" (1859, p. 61). To conclude, the presence of a human supralaryngeal vocal tract in a fossil hominid is an index for the brain mechanisms that allow voluntary control of speech and execute the rapid motor commands that are necessary for human speech. The earliest stages of specialization for human speech could have been built up on a general primate base *if* voluntary neural control of vocalization was in place. The initial increase in fitness from more efficient vocal communication might be derived without additional neural modifications for speech perception beyond those occurring in present-day apes. Studies with chimpanzees show that they can perceive human speech by using formant transitions and fundamental frequency contours (Savage-Rumbaugh et al., 1986). Therefore, the initial contribution to biological fitness of the human supralaryngeal vocal tract could have been to produce more distinct, unnasalized, quantal sounds without the increase in data rate that follows from formant frequency encoding. However, we know that our speech, the end point (so far), is encoded for rapid data transmission. The presence of a functionally modern human vocal tract 125,000 years ago and its subsequent retention and elaboration are consistent with the presence in this period of brain mechanisms allowing automatized speech motor activity, vocal tract normalization, and the decoding of encoded speech.

3

A Thoroughly Modern Human Brain

Any theory concerning the evolution of language must account for the brain mechanisms that underlie human linguistic ability. The theory proposed here involves two stages: the evolution of lateralized brain mechanisms in the archaic hominids who are our remote ancestors, followed by the evolution of the brain mechanisms that make possible rapid encoded human speech and syntax, as well as some aspects of cognition.

The theory follows from the circuit model of the brain discussed in Chapter 1. The human brain consists of many special-purpose mechanisms, which work together in different circuits to accomplish various activities. A particular mechanism, through complex connections to different parts of the brain, can participate in different aspects of behavior that logically seem to be unconnected. For example, Norman Geschwind showed in 1965 that a particular "motor control" area also regulates emotional responses, particularly ones involving fear and anger. The different behavioral roles reflect the evolutionary history of the brain (MacLean, 1985); mechanisms that evolved to serve one function were adapted to other ends. The evolutionary scenario that I propose is based on this Darwinian model; it therefore is not novel except that it attempts to account for the evolution of one aspect of language—syntax—that influential linguists such as Noam Chomsky believe could not have evolved by Darwinian processes (Chomsky, 1972, p. 97; 1980a, p. 3; 1980b, p. 182).

Stage 1: Brain Lateralization

The initial stage in the evolution of the neural bases of human language appears to involve lateralized mechanisms for manual motor control. Right-handed people, who constitute about 90 percent of the population, consistently use their right hand to perform precise manipulations. The left hemisphere of their brain, which controls the right hand, also controls the production of speech. The situation is often reversed for left-handed people—their right hemisphere controls speech.

The linkage between handedness and speech puzzled scientists for many years until Doreen Kimura, a neurologist who studies speech production, proposed that the lateralized brain mechanisms that regulate the production of speech evolved from mechanisms that initially were elaborated to facilitate skilled hand movements. Kimura notes that the skilled manual acts that are necessary for using and making tools

> requires the asymmetric use of the two arms, and in modern man this asymmetry is systematic. One hand, usually the left, acts as the stable balancing hand; the other, the right, acts as the moving hand in such acts as chopping, for example. When only one hand is needed, it is generally the right that is used. It seems not too farfetched to suppose that cerebral asymmetry of function developed in conjunction with the asymmetric activity of the two limbs during tool use, the left hemisphere, for reasons uncertain, becoming the hemisphere specialized for precise limb positioning. When a gestural system [for language] was employed, therefore it would presumably also be controlled primarily from the left hemisphere. If speech were indeed a later development, it would be reasonable to suppose that it would also come under the direction of the hemisphere already well developed for precise motor control. (1979, p. 203).

Kimura (1979) built on observations made early in this century (Liepmann, 1908). She showed that subjects who had suffered left-hemisphere brain damage had difficulty performing tasks involving coordinated novel movements of the fingers and thumb of one hand. In contrast, no decrements occurred in tasks involving simple or stereotyped movements.

Previous theories on the neural bases of human language, such as Eric Lenneberg's (1967), stressed the uniqueness of the lateralized human brain. However, recent data show that lateralization for handedness is not limited to humans; other primates show hand preferences under natural conditions (see Lieberman, 1984, pp. 66–70; MacNeilage, Studdert-Kennedy, and Lindblom, 1987, for reviews). One hand tends to be used by the animal in tasks that require some degree of precision. Although the brains of nonhuman primates are not as lateralized as those of modern humans, the difference is a matter of degree (Geschwind and Behan, 1984). No "special" evolutionary processes need have been involved in human brain lateralization (Lieberman, 1984, p. 69; MacNeilage, 1987; MacNeilage, Studdert-Kennedy, and Lindblom, 1987). Darwinian natural selection that gradually enhanced preexisting lateralization could have yielded the human condition, lateralized neocortical brain mechanisms that allowed voluntary vocal communication.

Stage 2: Speech, Syntax, and Some Aspects of Thinking

Human syntactic ability is ultimately related to the motor control of simple animals because there is continuity in the evolution of the neural mechanisms for motor control (Brooks, 1986). However, the abilities of human beings are far superior to those of even closely related, clever animals such as chimpanzees. But what factor or factors could have played a part in the evolution of modern human beings, who differ so profoundly from other animals? Clearly, upright posture, toolmaking, and social organization shaped the particular course of hominid evolution. However, given the anatomical and cultural differences that exist between early hominids such as the australopithecines (who walked upright, probably made tools, and probably had fairly complex social structures)[1] and anatomically modern *Homo sapiens*, some other factor must have been involved in the evolution of the brain mechanisms that underlie modern human language and thought.

I propose that natural selection to enhance faster and more reliable communication is responsible for the second stage of the evolution of these mechanisms—the evolution of the modern human brain. Communication places the heaviest functional load on "circuitry" for both electronic devices and brains. The transistors and solid-state devices that made digital computers a useful tool were first developed for communications systems. Indeed one can argue that the demands of communication preempt the highest levels of technology and organization of a culture, whether couriers on horses or lasers and fiberoptic bundles are the means employed. In short, evolution for efficient, rapid communication resulted in a brain that has extremely efficient information-processing devices that enhance our ability to use syntax. These brain mechanisms also may be the key to human cognitive ability. As many scholars have noted, human language is creative; its rule-governed syntax and morphology allow us to express "new" sentences that describe novel situations or convey novel thoughts. The key to enhanced cognitive ability likewise seems to be our ability to apply prior knowledge and "rules" or principles to new problems.

Syntax, Encoding, and Speed

Human languages almost always use syntactic rules that order the words of a sentence to convey meaningful distinctions. The meanings of words also can be systematically modified by *morphemes*, sounds that convey meaning. For example, the past tense of regular English verbs is conveyed by the morpheme transcribed by the letters *ed*—*walked, laughed,* and so on. The rules that we follow to convey meaning by modifying words constitute *morphology,* but the boundary between syntax and morphology is not hard and fast. A speaker of English, for example, must coordinate the morphemes that convey number in verbs and nouns by means of syntactic rules that operate across the entire sentence: *The boy is here* versus *The boys are here.* Syntactic rules differ from one language to the next; adjectives that precede a noun in English follow the noun in French: *the blue house* versus *la maison bleu.* However, virtually all human languages convey distinctions of meaning by means of syntax.[2] The ability to communicate novel ideas by means of novel sentences is one of the most powerful aspects of human language. Syntax is clearly species-specific. No other living animals, including language-trained apes, have been able to master anything but the very simplest syntactic rules. The syntactic abilities of the most proficient ASL-using chimpanzee are surpassed by human three-year-olds; by age four or five humans can produce an infinite number of new sentences.

Syntax also increases the speed of vocal communication by allowing us to "encode" several thoughts into the time frame that otherwise would transmit one simple thought. We could communicate by using sentences such as *The boy is small; the boy has a hat; the hat is red.* But we don't, because we can encode the same thoughts in the sentence *The small boy has a red hat,* which we can say in about the same time as a one-thought sentence. Human syntax, like human speech, allows us to communicate more information in a given time, and thus also allows us to circumvent some of the limitations of short-term memory. We do not have to keep track of the common referents of the three simple sentences, which we have to store in short-term memory. If we mean to communicate the facts that the *same* boy is small and has a hat, and that the hat is red, there is less chance for confusion when we use the sentence *The small boy has a red hat.*

A parallel can thus be drawn between the formant frequency encoding of human speech, which allows us to overcome the temporal resolution of the mammalian auditory system (the number of sounds that we can identify per second), and the encoding of syntax, which allows us to overcome memory limits. Both encoding schemes allow humans to communicate complex thoughts rapidly despite biological restrictions that derive from our primate heritage. In short, I propose that rapid, precise vocal communication was the engine that produced the modern human brain.

Speech Motor Control

The evolutionary basis for the brain mechanisms that underlie human syntactic ability likewise appears to be rapid vocal communication. The brain mechanisms that allow the production of the extremely precise complex muscular maneuvers of speech may have provided the preadaptive basis for rule-governed syntax (Lieberman, 1984, 1985). Motor control is always a difficult task. For example, the circuits and computer programs required to direct industrial robots are extremely complex. The explicit instructions necessary to command a robot must be expressed as complex rules that are similar in form to the rules of the grammar of human language. If, for example, you want to design a "smart" robot that will attach bumpers to automobile bodies, you have to provide a set of instructions that will first allow the robot to recognize the types of car bodies that take different bumpers. The robot must then select the appropriate bumper and attach it in the appropriate place. The rules that the robot needs to follow are *context-dependent* rules that do not conceptually differ from the syntactic rules that we use when we use the plural verb *are* that "agrees" with the plural subject in the sentence *The boys are playing.* Context-dependent rules regulate most aspects of human life. For example, you normally would not telephone neighbors between midnight and seven A.M., but if their house was burning you would.

The production of speech appears to be the most difficult motor control task that humans perform. The instructions that the brain must transmit to the muscles of the tongue and other speech organs have the complexity of the syntactic rules of human language and some aspects of rule-governed logic. Consider the automatized lip-rounding effects in the production of words such as *two.* When you say *two* you project your lips forward and simultaneously purse them 100 milliseconds before uttering the lip-rounded vowel [u] of *two.* In contrast, the vowel [i] of *tea* is not lip-rounded and you don't round your lips at all. You automatically carry out the context-dependent rule:

$$\alpha \text{ round lips } / \text{ [t] } \alpha \text{ vowel}$$

The symbol α conveys the state of lip-rounding. If the vowel is rounded (+ rounded), α has the value + rounded for the consonant. If the vowel is not rounded, α has a − value for the consonant. Therefore, the rule "reads": round the consonant so that it agrees in rounding with the vowel that follows it. It is no different in form from similar linguistic rules presented by Chomsky and Halle (1968). Rules of this form, coupled with sufficient memory, are sufficient to account for the full complexity of languages such as English (Gazdar, 1981).

Relevant Neurophysiologic Data and Theories

Unfortunately, we have no direct knowledge of the brains of the fossil hominids who represent intermediate stages in the evolution of anatomically modern *Homo sapiens*. However, we can study and compare the brains and behavior of living human beings with those of other living animals. In approaching the relation between the brain and human language and cognition, we must necessarily focus on humans because no other animal displays the full range of human language or cognition. However, comparative studies are useful because the brain bases of human language appear to involve the interplay of mechanisms that also function in other aspects of behavior. In some cases comparative studies reveal clear parallels between humans and other animals in the way that certain parts of the brain work. Other studies reveal differences that point out the ways in which our brains differ from those of closely related species. These unique aspects of the human brain are the ones for whose evolution we must account.

Broca's Aphasia

Over the past century a fund of knowledge has accumulated from data based on the "experiments in nature" that we briefly noted in Chapter 1. Extensive damage to the dominant hemisphere of the human brain in and near Broca's area can result in aphasia, a complex of speech and language deficits. In contrast, in monkeys massive lesions in or near the homologue of Broca's area or other parts of the neocortex have no effect on vocalization.[3] Their vocalizations are instead controlled by the cingulate cortex, the "old" motor cortex that evolved with the earliest mammals, the basal ganglia, and midbrain structures.

Monkeys and apes, lacking functional neocortical vocal control, likewise cannot produce the muscular maneuvers that are necessary to produce human speech. Broca's area and the circuits that connect it to other parts of the brain appear to be one of the unique characteristics of the human brain. The traditional view of Broca's aphasia[4] is that damage localized to Broca's area will result in these deficits, whereas damage to any other part of the brain will not. This belief is reflected in popularized accounts of how the human brain works, and in the supposition of many linguists that human beings have a specific, localized "language organ" (Chomsky, 1975, 1976, 1980a, 1980b, 1986). However, that supposition is erroneous. Donald Stuss and D. Frank Benson, in their comprehensive study of the anatomy and function of the frontal lobes of the human neocortex, note that damage to "the Broca area alone or its immediate surroundings (a lesion that could be called 'little Broca') is insufficient to produce the full syndrome of Broca's aphasia, at least not permanently . . . The full, permanent syndrome (big Broca) invariably indicates larger dominant hemisphere destruction . . . including the area of Broca but extending deep into the insula and adjacent white matter and possibly including basal ganglia" (1986, p. 161). The damage pattern that produces Broca's aphasia interrupts the *circuits* between Broca's area and other parts of the brain. The cortical and subcortical pathways connecting Broca's area to the parts of the brain that directly control muscles and to the prefrontal cortex can be interrupted by the massive tissue destruction that causes permanent aphasia. In fact subcortical damage that disrupts the connections from Broca's area but *leaves it intact* can result in aphasia (Naeser et al., 1982; Benson and Geschwind, 1985, pp. 206–207; Alexander, Naeser, and Palumbo, 1987; Metter et al., 1989).

Broca's aphasia also produces other motor deficits. The right hands, arms, or legs of about 80 percent of all Broca's patients are weakened. The face and arms are generally more affected than the legs, a result consistent with the phylogenetic history of Broca's area and Kimura's (1979) lateralization theory.

The most obvious language deficit of Broca's aphasia—impaired speech—is also motoric. Spontaneous speech is often absent; speech when it does occur is hesitant, labored, and distorted. The timing of the coordinated motor activity that is necessary to produce the sounds of speech is impeded. Many of the acoustic cues that differentiate speech sounds involve extremely precise control of the timing between different muscles. For example, the distinction between the English *unvoiced stop consonants* ([p], [t], and [k]) and *voiced stop consonants* ([b], [d], and [g]) is determined primarily by the timing of the phonation produced by the larynx relative to the release of the obstruction of the supralaryngeal vocal tract by the lips or tongue. This distinction, which has been termed *voice-onset time (VOT)*, constitutes one of the primary perceptual cues that differentiate these classes of sounds. Phonation starts soon after the release of the supralaryngeal constriction for the voiced consonants. Normal speakers maintain a separation of the VOTs of the two classes of consonants. One of the speech deficits associated with Broca's aphasia involves a loss of the control of VOT timing—the VOTs of the two classes of sounds merge (Blumstein et al., 1980). Broca's aphasics also have difficulty with nasal consonants such as [m] and [n], which involve synchronizing the opening of the nose and mouth (see Blumstein, 1981; and Caplan, 1987, for reviews of studies of aphasic speech).

Although lesions that are limited to Broca's area do not cause permanent language deficits, they can result in *aphemia*.[5] The victim is initially mute but then recovers. One interesting motor deficit that is significant in light of Kimura's preadaptive theory (that the brain mechanisms for speech motor control evolved from ones adapted to precise manual activity) is that lesions localized to Broca's area often result in "residual clumsiness and sensory disturbance of the right hand [which] may persist, especially involving the thumb and forefinger" (Stuss and Benson, 1986, p. 164).

Lesions centered in the anterior cingulate cortex and supplementary motor areas of the brain (see Figure 1–3) can also cause mutism in humans, similar to the effects produced in monkeys (Sutton and Jurgens, 1988). Recent studies that make use of computed-tomographic (CT) scan data, which produce x-ray images of the damage to a living patient's brain with a precision hitherto impossible, show that circuits involving subcortical brain mechanisms play an important role in human speech and language. Damage to subcortical structures such as the internal capsule (the bundle of nerve fibers that connect the neocortex to the midbrain) and the putamen (one of the structures of the basal ganglia; see Figure 3–1) can yield impaired speech production and *agrammatism* (deficits related to syntax) similar to that of the classic aphasias, as well as other cognitive deficits (Naeser et al., 1982; Alexander, Naeser, and Palumbo, 1987; Katz, Alexander, and Mandell, 1987; Metter et al., 1987, 1989). We will return to these experiments in nature and the relevant neuroanatomy, but we should first explain what we mean when we say that brain damage can cause agrammatism.

Agrammatism is one of the characterizing features of Broca's aphasia, although not every Broca's patient is agrammatic, nor does every agrammatic aphasic necessarily have speech production deficits.

The speech of Broca's aphasics is sparse and in some cases is simply limited to the repetition of single words or phrases. The "little" function words and morphemes that signal syntactic relationships such as *the, is, by,* and *under,* as well as the morphemes that indicate past tense and plural in English, may be omitted. Stuss and Benson note that although verbal output is the most obvious problem associated with agrammatism, it is not the only one: "the patient with agrammatism has the same difficulty, apparently to the same degree, in comprehending . . . syntactical structures" (1986, p. 169). The syntactic deficits of Broca's aphasia are not limited to speech. "Most Broca aphasics have difficulty understanding written material . . . particularly the words important for syntactical relationships . . . Comprehension of spoken language and comprehension of written material are defective in a similar manner. Thus most Broca aphasics comprehend some written material, usually nouns and action verbs (e.g., they may interpret a headline correctly but cannot understand the more detailed article)" (Stuss and Benson, 1986, p. 162).

Formal tests of Broca's aphasics show that comprehension is poor when only syntactic information is present (Zurif and Caramazza, 1976; Zurif and Blumstein, 1978). Experiments in

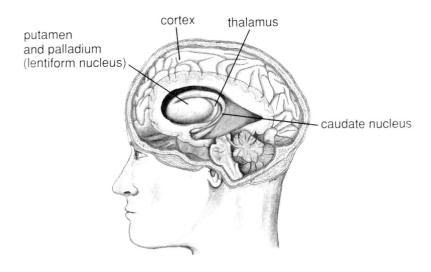

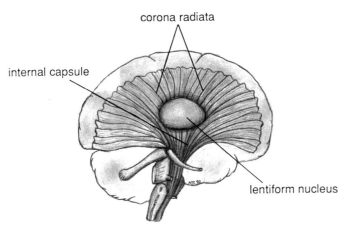

Figure 3–1. The basal ganglia are located deep in the cerebellum. The lentiform nucleus, which itself consists of two of the main basal ganglia structures, the putamen and palladium, is cradled in the set of descending nerves that converge to form the internal capsule and snake their way down through the basal ganglia. The caudate nucleus is another major basal ganglia component. The putamen, palladium, and caudate nucleus connect to the thalamus through complex circuits.

which agrammatic aphasics were asked to choose which of two pictures correctly captured the meaning of a tape-recorded sentence showed that they could not comprehend a sentence such as *The boy that the girl is chasing is tall.* In contrast, they could comprehend sentences such as *The apple the boy is eating is red.* They could make use of pragmatic cues and real-world knowledge to determine the meaning—apples are often red, and apples can't eat people. The grammatical abilities of Broca's aphasics are also impaired, and they have difficulty in grouping the words of a sentence in terms of its syntactic structure (Zurif, Caramazza, and Myerson, 1972). The syntactic structure of the sentence *The dog chases a cat* can be exemplified by the accompanying tree diagram. Normal subjects usually group the words of this sentence

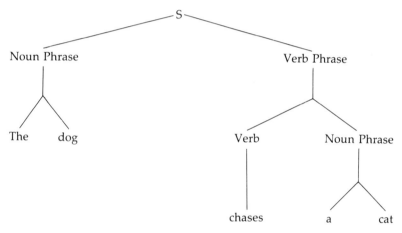

in terms of the sentence's syntactic structure, as:

(The dog)   (chases)   (a cat)

In contrast, Broca's aphasics grouped the words without regard to the syntactic structure, as:

(The)   (dog chases)   (a)   (cat)

Broca's aphasics also have high error rates when they are asked to decide whether sentences are grammatical or not. Although one study claimed that agrammatic Broca's aphasics can judge whether sentences were grammatical even though they could not comprehend the sentences (Linebarger, Schwartz, and Saffran, 1983), the data from the experiment in fact show extreme deficits in aphasic subjects' grammaticality decisions in comparison with normal controls. Similar effects are apparent in the data of Beverly Wulfeck (1988) and Shari Baum (1988), who also studied agrammatic aphasic subjects.[6] Baum's data and those of several independent experiments (Milberg, Blumstein, and Dworetzky, 1985; Tyler, 1985, 1986) show that the comprehension deficits of Broca's aphasics derive from an impairment in their ability to apply the rules of syntax automatically as they listen to the stream of speech. The rules seem to be stored, but they cannot be rapidly accessed and used in an automatic manner.

Wernicke's Aphasia

Wernicke's aphasia may derive from damage to this posterior area of the brain and circuits to and from it (see Figure 1–3). Wernicke's aphasics have difficulty comprehending speech. They also have difficulty naming objects or pictures, substitute inappropriate words for others, and coin new, meaningless words. Their speech often sounds fluent but meaningless. Wernicke's aphasia may involve the impairment of the brain's ability to integrate the acoustic cues that human listeners use to perceive speech with other aspects of language (Blumstein, 1981). Wernicke's aphasics do not show the deficits in the "automatic" comprehension of syntax observed for Broca's aphasia; instead they behave like normal controls in tests that involve *on-line,* that is, immediate, comprehension of syntax (Milberg, Blumstein, and Dworetzky, 1985; Tyler, 1985, 1986).

Sign Language Aphasia

Deaf people whose native language is American Sign Language (ASL) also can suffer aphasia. ASL aphasia, like that of speakers of vocal languages, usually results from damage to the left hemisphere of the brain. Ursula Bellugi, Howard Poizner, and Edward Klima (1983) studied three deaf signers who had suffered left-hemisphere strokes with different patterns of damage and a control group of age-matched deaf signers.[7] They administered a battery of formal language tests to each subject: tests of the morphological and syntactic structure of ASL, tests of the subjects' ability to perform hand and arm movements, and tests of their ability to perceive nonlinguistic visual signals. All three aphasic subjects had a relatively intact capacity to process nonlanguage visual-spatial relationships.

Two of the deaf aphasic subjects had deficits in syntax, which in ASL is expressed by the position and movement of the hands relative to the face and body, rather than by word order. For example, the hand signing a verb must move either from the location in which the subject was signed to the place where the object was signed, or vice versa, depending on the verb class. Although the means used to convey syntactic relations differ so drastically between ASL and spoken languages, similar effects occurred. Subject P. D., who had left-hemisphere subcortical damage near and behind Broca's area, was agrammatic. Subject G. D., who had suffered massive damage to Broca's area and most of the anterior frontal lobe, was unable to produce fluent sign language; her signing was agrammatic. Her comprehension of individual signs was good. Her deficits also resemble those of hearing Broca's agrammatic aphasics. In contrast, subject K. L. had damage in the left parietal lobe that extended subcortically into the posterior regions of the brain. The pattern of brain damage and of language deficits in many ways resembled that of Wernicke's aphasia. Her signing was fluent and grammatical. Her errors involved lexical substitutions.

Cognitive Deficits of Aphasia

Kurt Goldstein (1948), who studied aphasia over a period of fifty years, stressed the general cognitive deficits that accompany the specific linguistic deficits of aphasia. Goldstein differentiated between two "attitudes . . . the concrete and the abstract":

> in the concrete attitude we are given over passively and bound to the immediate experience of unique objects or situations. For instance, we act concretely when we enter a room in darkness and push the button for light. If, however, we desist from push-

ing the button, reflecting that by pushing the button we might awaken someone asleep in the room, then we are acting abstractly. We transcend the immediately given specific aspect of sense impressions . . . and consider the situation from a conceptual point of view and react accordingly. Our actions are determined not so much by the objects about us as by the way we think about them; the individual thing becomes a mere accidental example or representative of a "category." Therefore, we also call this attitude the categorical or conceptual attitude. The abstract attitude is basic for the following potentialities:

1. Assuming a mental set voluntarily, taking initiative, even beginning a performance on demand.
2. Shifting voluntarily from one aspect of a situation to another, making a choice.
3. Keeping in mind simultaneously various aspects of a situation; reacting to two stimuli which do not belong intrinsically together.
4. Grasping the essential of a given whole, breaking up a given whole into parts, isolating them voluntarily, and combining them in wholes.
5. Abstracting common properties, planning ahead ideationally, assuming an attitude toward the "merely possible," and thinking or performing symbolically.
6. Detaching the ego from the outer world.   (1948, p. 6)

Recent studies confirm Goldstein's view that the brain damage that causes aphasia also can result in cognitive deficits. Although standard tests of intelligence may show no decrement in overall intelligence, more specialized tests demonstrate that aphasic patients have considerable difficulty in performing tasks that involve keeping track of and applying different abstract concepts, translating specific facts into appropriate action, handling simultaneous sources of information, relating isolated details, and failing to grasp the key element of a problem (Stuss and Benson, 1986, pp. 194–203).

Subcortical Brain Damage and Disease

In the past two decades CT scans have made it possible to determine the location and extent of brain damage in individual living patients. Still more recently, positron emission tomography (PET) scans have allowed researchers to monitor the metabolic activity of various parts of the brain while a subject performs different linguistic or cognitive tasks. Studies using these technologies have confirmed what researchers suspected more than half a century ago but could not prove (Marie, 1926–1928): the role of subcortical structures in aphasia.

Michael Alexander, Margaret Naeser, and Carol Palumbo (1987) reviewed the language impairments noted in nineteen cases of aphasia, ranging from fairly mild disorders in the ability to recall words, to "global aphasia," in which the patient produced very limited incomprehensible speech, was unable to comprehend syntax, and was able to produce only hoarse, *dysarthric*, speech. Dysarthric speech results from a loss of motor control of the respiratory system and larynx; voice quality becomes hoarse and "breathy." Lesions in the basal ganglia and the nerve connections that run down from the neocortex were noted for all nineteen cases. In general, severe language deficits occurred in patients who had suffered the most extensive subcortical brain damage. These patients also suffered paralysis of their dominant right hand.

Alexander and his associates ascribe the speech and language deficits that result from subcortical damage to disruptions in "all the descending outputs from Broca's area and inferior motor cortex" (1987, p. 985)—the areas of the cortex that directly control the muscles involved in speech production. They note that subcortical circuits are probably redundant, so that small lesions have little effect; speech production deficits increase as more connections are cut. However, this explanation cannot account for all the aphasic deficits that characterize subcortical lesions. *Ascending* circuits, that is, connections back to the neocortex, also must be disrupted by the subcortical damage, since agrammatism similar to that of classic Broca's aphasia also occurred for five of the nineteen patients studied. The comprehension of syntax clearly is not a motoric activity. Therefore, aphasia resulting from subcortical damage demonstrates that human linguistic ability involves subcortical brain circuits that ascend back to the neocortex.

Recent studies using CT and PET scans of the same aphasic

patients bear on this issue. Jeffrey Metter and his associates (1989) examined twenty-eight aphasic patients whose behavioral symptoms fit the categories of Broca's, Wernicke's, or conduction aphasia (conduction aphasia is said to involve lesions in a cortical pathway between Wernicke's and Broca's area; Caplan, 1987, p. 56). The CT scans of the Broca's patients showed deep subcortical damage that included the internal capsule and parts of the basal ganglia. The PET scans showed that Broca's patients had vastly reduced metabolic activity in the left prefrontal cortex and Broca's region. Wernicke's patients had mild to moderate losses. Conduction aphasics did not differ from normal controls. One of the Broca's patients had no damage to Broca's area itself but nonetheless showed low metabolic activity in Broca's region and in the left prefrontal cortex. Previous studies by Metter and his colleagues (1987) showed a strong correlation between prefrontal and Broca's region metabolic activity and "functional motor loss of the arms and legs, as well as spontaneous speech and writing [and] in normal subjects, a strong correlation between prefrontal cortex and decision making" (1989, p. 31). They conclude that the behavioral deficits of Broca's aphasia—general "difficulty in motor sequencing and executing motor speech tasks" and "the presence of language comprehension abnormalities"—derive from damage to circuits to the prefrontal cortex.[8]

Diseases such as Parkinson's disease (PD) and progressive supranuclear palsy (PSP) show the effects of subcortical activity on cognition and language. These diseases spare the cortex but cause major damage to the basal ganglia, the subcortical parts of the brain that derive from our therapsid ancestors. The primary deficits of subcortical disease are motoric: tremors, rigidity, and repeated movement patterns. Alterations of personality also occur; mood can vary from depression to manic euphoria. Subcortical diseases can also cause cognitive deficits. In extreme form the deficits associated with these subcortical diseases constitute a "dementia," sufficient to impair the victim's ability to work and interact with other people (Albert, Feldman, and Willis, 1974; Pirozzolo et al., 1982; Cummings and Benson, 1984; D'Antonia et al., 1985).[9]

Progressive supranuclear palsy presents the most convincing case for profound subcortical dementia. The disease is characterized by "massive neuronal degeneration in the basal ganglia" (Pillon et al., 1986). Patients develop progressively slow slurred speech; they walk with shuffling steps. They have progressively greater difficulty in making arithmetic calculations. They need a longer than normal time to remember words and ideas and show a diminished ability to "manipulate acquired knowledge" (Albert, Feldman, and Willis, 1974).

In 1974 Martin Albert, a neurologist in the Aphasia Research Unit of the Boston Veterans Administration hospital, noted that the cognitive deficits associated with PSP are similar to those that occur when the frontal regions of the neocortex are damaged (Benson and Geschwind, 1972). Albert and his colleagues (Albert, Feldman, and Willis, 1974) looked at the known anatomy of the brain and argued that these deficits result from damage to subcortical circuits connecting various parts of the cortex to the frontal neocortex. As we shall see later in this chapter, the evolution of the brain follows the typical opportunistic "logic" of evolution. As the new cortex was added to the old reptilian brain, the basal ganglia took on new functions. Instead of simply directing signals down to the spinal column, they were modified to connect various regions of the new cortex with other parts of the brain.

Recent studies support this analysis. PET scans comparing normal subjects with PSP patients show that the latter have less metabolic activity in the prefrontal cortex. The neurons of the prefrontal cortex are not as active because they have been disconnected from the other parts of the brain that normally transmit information to them. The destruction by PSP of the circuits through the basal ganglia that stimulate the prefrontal cortex is responsible for the reduced activity and concomitant cognitive deficits (D'Antonia et al., 1985).

Parkinson's disease (PD) also affects subcortical basal ganglia pathways. The substantia nigra, a part of the midbrain that produces the hormone dopamine, is affected. The substantia nigra connect to and from the rest of the basal ganglia and other parts of the brain. The basal ganglia cannot function normally when the production of dopamine is deficient. L-DOPA, which supplies dopamine to the brain, is often administered to mitigate the symptoms of PD. The primary symptoms of Parkinson's disease are profound motor system deficits. Subjects have difficulty in executing voluntary movements; they claim, for example, that "my hands won't do what I tell them to." However, it is becoming clear that PD "may also impair the organisation of actions at a higher level, that is, in the decision making or planning level of skill" (Flowers and Robertson, 1985, p. 527).

Parkinson's disease subjects who have moderate cognitive deficits also have difficulty comprehending sentences that have somewhat complex syntax. A study of forty-four PD subjects with mild to moderate nonspeech motor control problems showed deficits in the comprehension of syntax and cognitive losses similar to those noted by Goldstein (1948) when aphasic patients lose the capacity for the "abstract attitude" (Lieberman et al., 1990). Cognitive capacity was evaluated by means of standard neurological tests. None of the subjects was judged to be demented, but the cognitive capacity of eight appeared to have diminished over the course of several years. All forty-four subjects were given the Rhode Island Test of Language Structure (RITLS), originally developed by Trygg and Elizabeth Engen (1983) to test the linguistic skills of hearing-impaired children. The RITLS assesses the extent to which subjects are able to use syntactic properties, such as word order and markers of the relationships between clauses in complex sentences in the understanding of sentences. The test presents a representative sample of the syntactic structures of English and simple sentences, such as *The man is old* and *The cat chased the dog*, as well as more complex ones, such as *The boy is small but strong* and *The dog was chased by the cat*. Because syntax is the focus of the test, vocabulary and morphology are tightly controlled: a small number of words judged to be understandable by young children is used repeatedly.

The RITLS sentences were read slowly and clearly, and the subjects were encouraged to ask the examiner to repeat the sentence as often as they liked in order to reduce the effect of any hearing or memory loss on comprehension. After hearing each sentence the subjects were shown three pictures and were asked to choose the picture that best characterized it. For example, for the sentence *Before the man washed the car, he put on boots* the choices were (1) a man putting on boots outside his house, (2) a man wearing boots washing a car, and (3) a man putting on boots next to a car, a bucket, and a running hose. Six of the eight subjects who had some loss in cognitive capacity showed systematic deficits in the comprehension of a moderately complex syntactic rule. One example of a syntactic category that they found difficult was "passive reversible" sentences, such as *The boy was hugged by the girl.* Only two of the thirty-six subjects who showed no signs of cognitive loss made similar errors. Since normal six-year-olds have only slight difficulties in comprehending the sentences of the RITLS, these deficits are meaningful. An independent study (Illes et al., 1988) showed similar decreases in the syntactic complexity of spontaneous speech of PD patients. They tended to use more nouns and verbs and fewer grammatical function words (such as prepositions).

Recent data show that the nature and pattern of speech production and syntax comprehension deficits of PD are very similar to those of Broca's aphasia (Lieberman et al., in preparation). Forty PD outpatients, twenty of whom were classified as mild and twenty as moderate in terms of motoric impairment, were studied. None was demented. On a battery of cognitive tests the subjects' responses were within normal ranges. In addition, neither the neurologist who spent significant amounts of time with them nor their families believed there had been any cognitive decline. All the PD subjects recited lists of monosyllabic English words that started with stop consonants. The RITLS was also administered to each subject, using a technique that allowed the experimenters to determine how much time each subject took to respond to each sentence.

Normal speakers maintain a separation of the VOTs of voiced ([b], [d], and [g]) versus unvoiced ([p], [t], and [k]) stop consonants. Ten of the PD subjects had an apparent VOT timing deficit and merged VOT for voiced and unvoiced stops. Eighteen percent of their VOTs overlapped, compared with the 4 percent VOT overlap that occurs for normal (non-PD) speakers under similar experimental conditions (Miller, Green, and Reeves, 1986). Thirty of the PD speakers maintained the VOT distinction; they had a 4 percent error rate and were statistically similar to normal controls. Some subjects who had VOT timing errors made few sentence comprehension errors, and vice versa. However, the average syntax comprehension error rate for the subjects who had VOT timing deficits was significantly greater than that of the subjects who preserved the VOT contrast. (The mean error rate for the ten subjects with a VOT deficit was 11.4 errors, SD = 8.96; the other thirty subjects' mean error rate was 4.65 errors, SD = 4.14.) The subjects with a VOT deficit also took significantly more time to comprehend the sentences in the RITLS. (Their mean response time was 2.73 seconds, SD = 1.12 seconds, whereas the other subjects' mean response time was 1.96 seconds, SD = 1.03 seconds.)

The association of VOT timing deficits and syntax comprehension deficits merits attention, since recent data on Broca's aphasics show that VOT timing deficits appear to be limited to patients with damage to subcortical structures (including the internal capsule, putamen, and caudate nucleus) and Broca's region (Baum et al., in press). The longer response times of the ten PD subjects who had a VOT deficit appear to reflect a need for increased processing time in order to apply their syntactic knowledge to the comprehension of a sentence. This result again is similar to the behavior of Broca's aphasics, whose performance improves when more time is available. Since these speech and syntax deficits occur in PD in the absence of any known cortical lesions, it is apparent that the basal ganglia must be included in the neural bases for human language.

A number of cognitive deficits have been observed in patients with PD. As Jeffrey Cummings and Frank Benson (1984) note, "a majority of patients with PD have measurable neuropsychologic deficits involving memory, abstraction, visual-spatial integration, and central processing time" (p. 874). Other researchers have found deficits in sorting tasks that involve planning (Flowers and Robertson, 1985) and in a variety of cognitive tasks, deficits similar to those associated with frontal lobe damage (Benson and Geschwind, 1972). Quantitative studies show cognitive deficits in PD even when the subjects have no obvious signs of dementia. Anne Taylor, J. A. Saint-Cyr, and A. E. Lang (1986) drew 40 subjects from a pool of 100 PD patients, excluding those who showed signs of dementia, drug-induced confusion, emotional problems, and uncontrollable motor problems, and matched them with 40 normal controls for age, sex, education, and IQ. A comprehensive battery of tests revealed no significant differences between PD and normal subjects in memory or visual tasks, except for one effect on a serial recall task. There was a highly significant difference between the PD and control groups in cognitive tasks that required the subject to derive a "meaning [that] lies in context-dependent contingencies"; the PD subjects showed "an inability to generate plans" (p. 871). Moreover, the severity of motor and mental dysfunction was correlated.

The Neurophysiology of Speech, Syntax, and Some Aspects of Cognition

The data discussed above show concurrent speech production, syntax, and particular cognitive deficits. The cognitive deficits involve the ability to integrate knowledge and procedures in "creative" tasks, ones that cannot be solved by the rote application of previously learned knowledge. Damage to subcortical pathways that connect the prefrontal cortex with other parts of the brain is the most probable cause of these deficits. The CT scan studies reviewed are consistent with this hypothesis.[10] Moreover, the PET scan data show that the activity of the prefrontal cortex is reduced in Broca's aphasia, with similar effects in basal ganglia disease—PSP (D'Antonia et al., 1985) and Parkinson's disease (Metter et al., 1984). These data are consistent with the neurophysiologic theory that follows.

The Prefrontal Cortex

According to Donald Stuss and Frank Benson, "The prefrontal cortex is the anatomical basis for the function of control . . .

The frontal lobes are imperative at the time a new activity is being learned and active control is required; after the activity has become routine, however, these activities can be handled by other brain areas, and frontal participation is no longer demanded" (1986, p. 244). Stuss and Benson develop their theory in the light of neuroanatomical and behavioral data of the past 100 years. Their theory is consistent with and, in effect, subsumes the theories developed by other neurophysiologists and neurologists (for example, Milner, 1964; Teuber, 1964; Luria, 1973; Shallice, 1978; and Fuster, 1980). They conclude:

> The control functions of the prefrontal cortex become more and more obvious . . . [in] language, memory, and cognition, reflecting the greater complexity of the functions.
> The ability to take the information extracted from other, higher brain systems, verbal and nonverbal, and to anticipate, select goals, experiment, modify, and otherwise act on this information to produce novel responses represents the ultimate mental activity; all available data indicate that these executive functions are prefrontal activities. (1986, p. 246)

The functions of the prefrontal cortex appear to be similar in nonhuman primates. Hans Markowitsch (1988), reviewing the data of about 500 studies, notes that the prefrontal cortex is involved "in a large number of behavioral acts, ranging from blood pressure, heart rate, and temperature regulation to personality factors, social interactions, and mnemonic achievements. The basis for these behavioral phenomena can be seen in the intimate anatomical connections of the primate prefrontal cortex as a whole with regions of the association and parasensory cortex, the basal ganglia . . . and with a number of brainstem structures" (1988, p. 135). The lesion, electrophysiologic, and anatomical studies that can be performed with nonhuman primates show that the prefrontal cortex is connected, directly or indirectly, with virtually every part of the brain. Lesions in the prefrontal cortex in nonhuman primates "lead to a limited capacity in dealing with tasks that demand either that complex interacting variables be kept in mind simultaneously or that behavioral strategies or tendencies be switched, replanned, or restructured" (1988, p. 100).

The prefrontal cortex has become proportionately larger in more advanced primates. As we saw in Chapter 1, the increased proportion of the frontal regions of the cortex relative to its total surface is apparent in the Brodmann (1908, 1909, 1912) maps of the brain. The human prefrontal cortex is twice as large as it would have been if we simply had a large ape brain (Deacon, 1984, 1988b).

The Basal Ganglia

Perhaps one of the reasons why we usually associate the "highest" cognitive and linguistic functions of *Homo sapiens* with the neocortex is that it is the product of the last few hundred million years of evolution. In contrast, the basal ganglia are part of our reptilian legacy. However, the basal ganglia of humans also show the signs of millions of years of evolution. As we noted earlier, they are about fourteen times larger than they would be if we had the brain of a large insectivore. Moreover, the structure of the basal ganglia of mammals differs profoundly from that of reptiles. There are correspondences, homologous structures, but their size, organization, and connections are profoundly different. In fact the basal ganglia of "advanced" mammals such as monkeys differ from the basal ganglia of "simple" mammals such as rodents. The process of evolution has not missed the old reptilian part of the human brain, and the basal ganglia play a direct part in the activities that set us apart from other animals—language and thought.[11]

As the top of Figure 3–1 shows, the human basal ganglia are a set of anatomical structures buried deep in the cerebellum. The key structures of the basal ganglia are the *caudate nucleus,* the *putamen,* the *palladium,* and two closely related structures, the *subthalamic nucleus* and *substantia nigra,* sited in the midbrain region that connects the cerebellum with the lower brain and spinal cord.[12] These midbrain structures are closely interconnected with the parts of the basal ganglia that are positioned within the cerebellum. The putamen and palladium together form a structure called the *lentiform nucleus.* The bottom of Figure 3–1 shows the *corona radiata,* the immense fanlike set of nerves that radiate from the inside of the cortex and connect with the spinal column—the connecting "cable" from the cortex to the midbrain and spinal column. Signals are transmitted down these nerves to neurons that, in turn, transmit signals to initiate activity in various parts of the body. Signals that enable the cortex to monitor motor activity and to sense heat, touch, and pain travel up to the cortex in these nerves. The bundle of fibers bunches up, forming the *internal capsule* that runs down, snaking through the basal ganglia. The lentiform nucleus is cradled in this set of nerve fibers; extensive damage to the region around the internal capsule will also damage the putamen and parts of the caudate nucleus.[13]

In the last few decades neurophysiologists have developed tracer techniques that allow them to determine the wiring pattern of the brain. A tracer is a chemical that flows along neural pathways, allowing specific patterns of connections to be seen. Some tracers are specific to a particular neurotransmitter; others infiltrate all the neurons that connect to a particular injection site. The tracer is injected into a part of a living animal's brain. After an interval of several days to allow the tracers to propagate down or up the pathways, the animal's brain is removed, stained, and sectioned for microscopic observation. This limits the technique to species that are not endangered and precludes tracer studies on human subjects. Tracer studies show that both reptiles and mammals have connections descending from the basal ganglia to the spinal column, where they trigger motor activity. However, in mammals half of the connections *ascend* to other structures of the brain and do not directly initiate motor activity. The mammalian basal ganglia also receive input from virtually all parts of the cortex (except for visual cortex inputs, which are minimal) and from other subcortical structures (Parent, 1986). In other words, the basal ganglia of a mammal can integrate information from one part of the cortex with other input and transmit a signal back up to different parts of the cortex.

Tracer studies show important differences between the basal ganglia of simple mammals such as rodents and those of monkeys. In rodents the putamen and caudate nucleus are not differentiated. This distinction is significant; recent studies show that in monkeys the motor cortex (which senses and directly

controls motor activity) sends signals exclusively to the putamen. The legs, arms, and face of the monkey are represented
in its putamen in a manner analogous to their representation
in the motor cortex—where stimulation of one particular location will affect motor activity in an arm, another location the
tongue, and so on. In contrast, the associative areas of the prefrontal, temporal, parietal, and cingulate cortices connect exclusively to the caudate nucleus. Distinct circuits connect the caudate nucleus and putamen with the substantia nigra, from
which different connections ascend back to the cortex. Figure
3–2 shows these two associational and motoric loops of the
basal ganglia. The segregation of speech and syntax deficits that
is often noted in PD and aphasia may reflect this anatomy.
Individual patients with speech production deficits do not necessarily have syntactic or cognitive deficits, because the basal
ganglia circuits that go through the substantia nigra from the
caudate nucleus and putamen are independent.

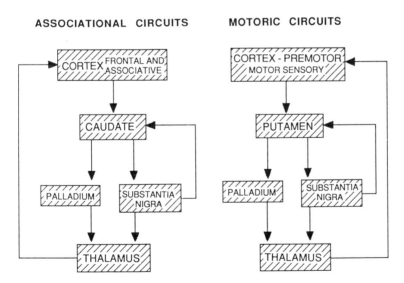

ASSOCIATIONAL CIRCUITS   MOTORIC CIRCUITS

Figure 3–2. The two neuronal loops of the primate basal ganglia. The caudate
nucleus is involved in various associative activities, the putamen in motoric
activity. Separate pathways go between these structures and different parts of
the substantia nigra. (Based on Delong, Crutcher, and Georopopoulos, 1983)

Broca's Area

Broca's area clearly is not *the* human language organ. It probably should be regarded as a multipurpose higher-level association area that is specialized to access subroutines for certain sequential operations. It appears to enter into different aspects of behavior through different circuits. It facilitates skilled hand-arm movements (Kimura, 1979). As Paul Broca first noted in 1861, it operates in the domain of speech. Broca's area also appears to access the rules of syntax. Through connections to the prefrontal cortex these automatized subroutines are applied to perform a novel manual act, utter a syllable, or comprehend the meaning of a sentence. The deficits of aphasia in this light are not exclusively linguistic; they are part of a more general problem.

Given the primitive state of our knowledge concerning the physiology of the brain, we cannot be certain as to precisely where or how the automatized motor subroutines or syntactic rules are stored or accessed. Memory, including that of the rules of syntax, probably involves some form of distributed neural network or networks (see Chapter 4). However, although distributed neural networks probably store this information, the memory traces must be accessed somewhere. The neurological data discussed above indicate that Broca's area is the most likely candidate for accessing these traces.

Some Stages in the Evolution of Syntax and Thinking

Given these data and theories, what can we say about the chronology and course of the evolution of human speech, syntax, and thought?

Stage 1: Brain Lateralization in Early Hominids

Since tool use is not a unique human attribute—chimpanzees regularly use one-handed tools (Boesch and Boesch, 1981, 1984; Goodall, 1986)—changes in the brain that enhanced the precise, one-handed use of tools would have contributed to fitness in the primates ancestral to hominids. Although some anthropologists claim that australopithecines did not make stone tools (Day, 1986), they probably used stone and wood tools, given the fact that even present-day chimpanzees use wood anvils and hammer stones to split nuts open (see Chapter 6). The fossil remains of *Homo erectus* and of all subsequent hominids are associated with evidence of stone tool manufacture. It is probable that the human pattern of lateralization for manual activity was established in *Homo erectus*. Once established, tool manufacture would itself enhance the biological fitness of brain mechanisms that yielded ever more precise manual dexterity.

Lateralization and Speech Perception

Brain lateralization is also associated with the perception of human speech (Liberman et al., 1967). The left hemisphere of the human brain is involved in processing speech sounds and nonspeech sound sequences that involve temporal-order decisions—deciding what sound occurs before or after another sound (Bradshaw and Nettleton, 1981). Some nonhuman primates also show brain lateralization effects in perceiving species-specific vocal calls (Peterson et al., 1978; Heffner and Heffner, 1984; Newman, 1988). There is a "right-ear" advantage—some of the sounds of human speech and certain nonhuman primate calls are more readily and accurately perceived when they are presented to the right ear, which is more directly connected to the left hemisphere of the brain. Nonhuman primates can perceive some of the sounds that convey meaning in human language, but it is not clear whether they can identify and differentiate the full range of human speech sounds (Kuhl, 1988). We need data that show how other animals interpret the full range of sounds of human speech. Lacking these data, theories on the evolution of lateralized brain mechanisms for speech perception remain speculative.[14]

Stage 2a: Voluntary Control of Speech

Vocal communication's initial contribution to biological fitness probably derives from the fact that it frees the hands. Since the earliest hominids could walk upright, their hands were free to use tools and carry burdens. Although manual sign language is an effective means of communication, it does involve using one's hands. Tool-using hominids would have enjoyed an advantage in most situations if they were capable of using vocal communication. Vocal communication is also more effective when the viewer is not close by or when light conditions are poor. About the only situation in which manual signals have some special utility is when silence is an advantage. Therefore, although scholars such as Gorden Hewes (1973) are probably correct in claiming that it may have played an important role in the earlier stages of hominid evolution, sign language was never the exclusive channel for human language.

A change in brain organization that allowed voluntary control of vocalization is the minimum condition for vocal communication. Broca's area and its connecting circuits allow human beings to use the automatized motor subroutines that control speech production. The homologue of Broca's area in nonhuman primates is the part of the lower precentral cortex that is the primary motor area for facial musculature. Dwight Sutton and Uwe Jurgens (1988) note that electrical stimulation of this area in squirrel monkeys (which have been studied in detail) yields isolated movements of the monkey's lips and tongue and some laryngeal activity but no complete vocalizations. Destruction of this part of the brain in squirrel monkeys and rhesus macaques has no effect on the calls or even on the performance of tasks that involve vocalizations. In contrast, stimulation of the anterior cingulate cortex elicits lip movements, phonation, and various calls in rhesus macaques; corresponding lesions in the cingulate cortex also interfere with calls. Stimulation in the supplementary motor association area (SMA) in the monkey's neocortex elicits phonation if the anterior cingulate is first stimulated. However, although the parts of the brain to which Broca's area is connected affect vocalization, the homologue of Broca's area is not involved.

The neocortical control of speech in humans undoubtedly involved both the addition of new parts and the rewiring of old circuits. It is likely that the enlargement of the prefrontal cortex reflects, in part, its role in speech production. The rewiring appears to involve the basal ganglia; data from recent comparative studies suggest that basal ganglia circuits may be the key to the unique brain bases of human speech and syntax. In studies involving electrical stimulation of human subjects, the experimenter applies low-amplitude electrical signals to the exposed cortex when a cooperating patient's brain has been exposed preparatory to surgery. George Ojemann (1983) found that stimulation along a set of arcs in and near Broca's area inhibited various aspects of speech and language. Terrence Deacon (1988a), using tracer techniques, has found similar pathways in the homologous parts of the macaque monkey cortex. These data, though inferential, suggest that the wiring of the human and nonhuman primate neocortex does not differ substantially. However, neurophysiologic and behavioral data show there *must* be a difference between the circuitry of the human and nonhuman primate brain with respect to the voluntary control of speech. "Experiments in nature" involving subcortical damage and disease and speech and language deficits point to the human basal ganglia as the missing link.

The new subcortical rewiring connects the prefrontal cortex with other neocortical brain structures as well as with older parts of the brain such as the cingulate cortex. These older brain mechanisms continue to function in human speech. The evolution of the basal ganglia probably did not stop with monkeys; in all likelihood basal ganglia circuits related to speech and language differentiate the human brain from other primate brains. Other specialized subcortical structures also may be implicated in the production of speech and probably have adapted to enhance speech production. Multiple redundant brain circuits seem to be the rule: redundancy is a virtue in the design of any machine that *has* to work, whether it is a brain or a Boeing 747 airliner. The process appears to be another example of the typical proximate logic of evolution (Mayr, 1982)—new mechanisms were added, old ones modified at minimum cost.

Stage 2b: Syntax

Brain mechanisms adapted to handle the complex sequential operations necessary for speech production would have no difficulty in handling the comparatively simple problems of syntax. Speech motor control involves rapidly invoking coordinated muscle activity patterns contingent on previous or future events. The rules that linguists devise to describe the syntax of human languages, though complex, are simple symbolic operations that can be reduced to substituting one symbol for another in a sequence of symbols (Bunge, 1984; Chomsky, 1986). The sequence of symbols that define a sentence determines the context in which one symbol is substituted for another. Similar rules describe speech motor control for even seemingly simple tasks, such as providing a sufficient supply of oxygen while we speak.

The muscular maneuvers that we perform when we fill our lungs with air in order to speak involve complex planning across the span of an entire sentence. Human speakers usually estimate the length of time it will take to produce all the words of the sentence they *intend* to speak when they inhale before speaking. The volume of air breathed in is proportional to the length of the intended sentence (Lieberman and Lieberman, 1973). First the duration of each word must be computed; then the durations of all the words that compose the sentence must be added to one another. If the total duration is too long, the sentence is broken into segments that correspond to major breaks in syntactic structure (Armstrong and Ward, 1926; Lieberman, 1967). The rule or algorithm for taking air into the lungs thus (1) operates across the frame of the entire sentence and (2) takes into account syntactic relationships. The "preprogramming" necessary to control the muscles that regulate the airflow and air pressure that determine the melody or intonation of the sentence also involves taking into account whether we are upright or reclining, jogging and conversing with a friend, the amount of fluid in our stomach, and linguistic factors—whether we are asking a yes/no question or emphasizing part of the sentence that we *intend* to speak (Atkinson, 1973; Bouhuys, 1974).

Preadaptation and Universal Grammar

The evolutionary model presented here does *not* deny the existence of brain mechanisms specialized for syntax. It does claim that speech motor control is the preadaptive basis, that is, the starting point. Once syntax became a factor in human communication, the selective advantages that it confers (increasing the speed of communication through encoding and circumventing short-term memory limitations) would have set the stage for natural selection that specifically enhanced these abilities—independently of speech motor control. Therefore, it is quite possible that some aspects of the brain mechanisms that regulate the syntax of human language have nothing to do with motor control. Natural selection over the past 150,000 to 200,000 years could have produced neural structures dedicated to syntax. Noam Chomsky's (1986) intuitions probably are, in part, correct—human beings undoubtedly have brain mechanisms that genetically code some aspects of the possible syntactic rules of all human languages. However, the biological "language organ" must conform to the general properties of biological organs—variation, maturation, and an evolutionary history. The issue will be resolved only when we know more about the physiology of the brain and about seemingly mundane matters such as the actual pattern of syntactic variation in the real day-to-day use of language.

When Did the Modern Human Brain Evolve?

The reorganization that makes the voluntary control of speech possible is one of the defining characteristics of the modern human brain. It undoubtedly had occurred 100,000 years ago in anatomically modern fossil hominids such as Jebel Qafzeh and Skhul V, who had modern human vocal tracts. It probably had already occurred in fossil hominids such as Broken Hill who may have lived 125,000 years ago, may have been ancestral to these fossils, and had a vocal tract that was almost completely modern. It may have occurred in the African fossil hominids (who have yet to be unearthed) ancestral to the early specimens of anatomically modern *Homo sapiens* (Stringer and Andrews, 1988). The dating of adaptations for syntax and abstract cognitive processes (which appear to derive from the human prefrontal cortex) is more difficult to determine. The archaeological evidence associated with Jebel Qafzeh and Skhul

V (see Chapter 6) is consistent with their possessing a fully modern human brain—one adapted for complex syntax and logic—but an earlier origin cannot be ruled out. Although much of the enlargement of the prefrontal cortex may derive from the specific contributions of speech and language to biological fitness, it enters into virtually all aspects of behavior. Therefore, any cognitive activity that enhanced biological fitness could have contributed to its development.

Neanderthal Brains

Although the presence of a modern human vocal tract is an index for the co-occurrence of the brain mechanisms that are necessary to run it, its absence in fossils such as the classic Neanderthals, who lived until about 35,000 years ago, does not demonstrate that they lacked these brain mechanisms. Since the brain mechanisms that underlie speech production appear to have evolved from ones that were adapted for manual motor control, some level of speech production ability probably existed in the archaic hominids who preceded anatomically modern *Homo sapiens*. Indeed, we can argue that some level of speech production ability *must* have existed then, because there would have been no selective advantage for the evolution of the human vocal tract *unless* speech communication was already in place. But the retention of a functionally nonhuman vocal tract in Neanderthal hominids for at least 70,000 years after the appearance of anatomically modern humans argues for different selective pressures—perhaps for less reliance on speech, since Neanderthal speech would, at minimum, be less distinct than human speech. Therefore, it is quite likely that Neanderthal brains were not as well adapted for speech—or syntax and other aspects of the "abstract attitude." Although many scholars (such as Terrence Deacon, 1990) have claimed that the brains of Neanderthals were the same as ours, their arguments always reduce to ones involving the overall size of the Neanderthal brain, which is in the human range. However, brain size in itself is not significant. It is impossible to determine the structure or the connections of a fossil brain by looking at its skull; we cannot say whether the Neanderthal brain had speech and language circuits equivalent to ours.[15]

Specialization and Generalization

Two systems of the human brain have been discussed throughout this chapter: (1) Broca's area and its connecting circuits and (2) the prefrontal cortex. A functional Broca's area (in which we must include connecting circuitry) is species-specific; it is *necessary* for speech production and syntax. Although it has evolutionary antecedents, only modern human beings can voluntarily access stored motor control programs for speech production. Only modern humans appear to possess the ability to produce or interpret sentences with complex syntax. Humans also appear to be the only animals who can handle complex logical propositions (see Chapter 6). However, Broca's area is clearly *not* the modular "language organ" postulated by linguistic theory. Its functional circuitry includes phylogenetically "old" brain mechanisms, and it retains its "older" motoric involvement with precise one-handed skills.

Moreover, the prefrontal cortex is also involved in both human language and thought. The circuits that connect it to Broca's area make language possible; indeed many of the activities traditionally ascribed to Broca's area may involve the activity of large parts of the prefrontal cortex and associated subcortical circuits. The enlargement and complexity of the human prefrontal cortex undoubtedly derive, in part, from the specific contributions of language to biological fitness. But the prefrontal cortex is also involved in all new, creative activity. It integrates information and appropriate motor responses, learns new responses, and derives general abstract principles. It is the brain's "think tank." We don't need it when everything is running smoothly and routinely, but it comes into action to solve problems and learn new responses.

The brain bases of human speech, syntax, and thought thus consist of a set of circuits that connect many old and new parts of the brain. The evolution of these circuits and mechanisms appears to follow from the Darwinian principles that generally account for the complex anatomy and brains of other animals.

4

The Brain's Dictionary

Until the 1960s linguists and philosophers believed that the ability to use words was *the* key to human language. For example, in 1964 Norman Geschwind claimed that other animals lacked neocortical brain circuits that supposedly were necessary for learning the meanings of words. Geschwind, who was one of the foremost neurologists of the century, was pursuing a reasonable goal; since other animals lacked human language, they must lack the neurological ability that he was attempting to identify.

Various experiments and observations have since demonstrated that other animals can acquire and use words. Experiments with chimpanzees in the late 1960s first demonstrated this for apes (Gardner and Gardner, 1969). Formal tests show that California sea lions and other marine mammals can learn the meanings of a limited set of words (Herman and Tavolga, 1980; Schusterman and Krieger, 1984). Parrots can be taught to respond to and produce a limited number of words (Pepperberg, 1981).[1] Many other animals such as dogs commonly respond to a few words; trained dogs have remarkable vocabularies (Warden and Warner, 1928). However, although many animals have lexical abilities, words are the essential feature of language. Without words language could not exist. Therefore, an evolutionary theory for language must account for the brain mechanisms that enable us to acquire and interpret words.

In the following pages I will argue that similar brain mechanisms, distributed neural networks, determine the lexical abilities of human beings and other animals. The main distinction between humans and other animals with regard to their lexicons appears to be quantitative. Chimpanzees and gorillas have the largest animal vocabularies; chimpanzees raised as though they were children using American Sign Language (ASL) acquire a maximum of about 200 words (Gardner and Gardner, 1969). In contrast, human children after age two and a half years go through a "naming explosion"; they learn so many words that it becomes impossible to ascertain the total number of words in their lexicon (L. Bloom, 1973; R. W. Brown, 1973). This is in itself an important distinction. Quantitative "computational" distinctions lead to qualitative *functional* differences for both computers and living beings. A ten-dollar calculator and a Cray supercomputer use similar computing schemes, but the nature of the problems they can solve differs qualitatively. The vastly greater number of words that people acquire allows them to think about and communicate a corresponding number of concepts. In other words, the enormous number of words that human beings know constitutes a unique aspect of human language and thought.

What Do Words Mean?

It is clear that words convey concepts rather than particular things or actions. This is the case for ASL-trained chimpanzees as well as for human beings. Although the chimpanzees' vocabulary is limited, their words clearly are not symbols that refer to particular objects. To a knowledgeable human, the American Sign Language (ASL) word *tree* does not refer to a particular tree. Formal tests in which color slides of new objects are displayed to ASL-using chimpanzees show that *tree*, to them, as to us, refers to a class of trees and treelike plants rather than to any specific tree (Gardner and Gardner, 1984). This aspect of the character of a word holds for all human languages. Jonathan Swift in *Gulliver's Travels* (1726) presented a parody of philosophic arguments concerning the imprecision of words in the Academy of Laputa's "Scheme for abolishing all Words whatsoever . . . since Words are only names for *Things*, it would be more convenient for all Men to carry about them, such *Things* as were necessary to express the particular Business they are to discourse on . . . which hath only this Inconvenience attending it: that if a Man's Business be very great, and of various Kinds, he must be obliged in Proportion to carry a greater Bundle of *Things* upon his Back, unless he can afford one or two strong Servants" (1970 [1726], p. 158). Even the simple word *shoe* can mean something very different to an automobile repairman and a fashion designer. It would also have meant something very different to an eighteenth-century farrier and Marie Antoinette. Words enhance communication and thinking precisely because they do not refer to specific things; they instead code concepts and reflect conceptual knowledge.

Formal Logic and Meaning

One of the unsolved problems of linguistics is explaining how we are able to sort out the different concepts that each of the words of a particular sentence could convey. For example, the meaning of the word *bank* in *The merchant deposited his money in the bank* is quite different from its meaning in *He fished all day on the side of the bank*. The most common words have the most meanings. But whereas humans have no difficulty in comprehending the meaning of the words of a sentence, computer systems designed to comprehend sentences find the task formidable.

Many linguists have attempted to use the mathematical procedures of formal logic to solve this problem. These attempts have not been very successful, because formal logic inherently involves making clear, categorical distinctions. Defining even a "simple" concrete word such as *table* is extremely difficult because some property must be associated with the word in an unambiguous way. Logicians usually define words by means of thought-exercises in which the world is divided into two different "sets" of objects or processes. In order to define the meaning of *table* we would have to establish a set of things that clearly are tables, and a set of things that never are tables. We would also have to specify a set of physical or functional properties that make something a table. This turns out to be impossible. We cannot even claim that tables have four legs: some tables have only one leg; some fold out from a wall and have none. Functional definitions do not work any better: tables support things, but so do beds or chairs. Words have fuzzy referents that cannot be characterized by the procedures of formal, mathematical logic. As Jacob Bronowski noted, "you cannot say anything about a table or chair which does not leave you open to the challenge, 'Well I am using this chair as a table.' Kids do it all the time, 'I am using this chair as a table, it is now a table or chair.' . . . And if that is true of 'table,' it is true of 'honor' and it is true of 'gravitation,' and it is true, of course, of 'mass' and 'energy' and everything else" (1978, p. 107). Bronowski rejected the relevance of formal logic to how we understand the meanings of words. And the philosopher Hilary Putnam (1981) has reached similar conclusions in a critique that demonstrates that it is *logically* impossible to define words by using the procedures of formal logic.

Dozens of psychological experiments show that words are constructs that associate meanings derived from experiences with a phonetic "name" and with each other (Lupker, 1984). When I consciously think of the word *bicycle* for more than a few seconds I almost always recall a scene in Brooklyn: the image of my father trotting alongside me on the Ocean Parkway bicycle path, steadying my bicycle on my first wobbly ride. For me, this early experience is indelibly associated with the word *bicycle*. When we think of a word we access both the name of the word and its meanings. The meanings derive from the experiences associated with the word, and the specific personal memories that underlie the word can come to mind. Although communication would be impossible if some common core of meaning did not exist, many of the difficulties that attend human life stem from the distinctions in meaning that we don't communicate, as Dostoevsky, Proust, and Conrad, among others, demonstrate.

Distributed Neural Networks

One of the problems that children have as they learn the meanings of words involves determining what aspect of life a word conveys. Is *doggy* the furry creature that licks you, or does it refer to all the creatures who move that way? What, then, is *kitty*? A child must discover the fuzzy boundaries of each word. She or he must somehow derive the concept coded by a sequence of sounds, the phonetic word, from a large number of experiences all of which vary somewhat. Various mechanisms have been proposed to account for how people and other animals learn from experience, but it is only in the last decade that a biologically plausible mechanism has become apparent—distributed neural networks.

A distributed neural network is a model of the presumed microstructure of the brain that inherently sets up patterns of association as it is exposed to different events or information. When exposed to the real-world experiences that constitute the "meaning" of a word, an appropriate distributed neural network would inherently associate this set of meanings with the phonetic "name" of the word. The word could then be accessed from the network by either its semantic or phonetic properties. In other words, the neural network would "recollect" the word either through the way that it sounds or by means of related meanings or events. For example, *collie* could be accessed by telling the distributed network to "think" about *poodles, sheep, Scotland,* words that begin with the letter *c*, or words that rhyme with *Molly*.

It is always difficult to decide on *the* characteristic that defines thinking. If thinking is the ability to derive abstract principles from empirical experience, then computer models of distributed neural networks think. The one developed by Terrance Sejnowski, Christof Koch, and Patricia Churchland (1988) can abstract the degree of curvature and orientation of a complex shape. The same network can learn to pronounce the words of English from normal text. Similar networks can readily derive the linguistic rule for forming the *ed* past tense of regular English verbs (Rumelhart, McClelland, and PDP Research Group, 1986). One of the interesting things about this last experiment was that the computer model's "behavior" was similar to that of a human child learning English. English has a set of "regular" verbs that take the written *ed* plural as well as "irregular" verbs such as *see* and *eat*. When children learn the regular English plural they almost always overgeneralize, forming words they have never heard, such as *eated*. Linguists correctly interpret this behavior as evidence that the child has learned the abstract linguistic principle or rule that underlies the utterances. The distributed neural network in this experiment overgeneralized the regular English past tense and formed words such as *seed* and *eated*.

Functional Cortical Maps

Few people would claim that the brain is simply one big amorphous distributed neural network. It is evident that the brains of even simple animals consist of specialized mechanisms, such as ones that are adapted to processing visual rather than auditory or olfactory input. It is also becoming evident that information that we commonly think of in a holistic manner, such as the picture of a tree, is processed and stored in a set of *functional maps* that are integrated at some higher level.

The commonest kind of map shows where things are in relation to other things. On a map of the United States, we can see that New York City is north of Philadelphia. But maps can also code other information with symbols for, say, the population of a city. We could also code the way in which the people who live in these cities voted in the 1988 presidential election. The maps that are formed in the cortex code different attributes of vision, hearing, and other sensations. Recent research shows that there are many cortical maps for each sensory mode (see Altman, 1987, for a review). Some visual maps, for example, code the color, size, or direction of movement of different objects. Other visual maps code more complex shape information that is of interest to the particular animal. The maps code attributes of the signal that are of interest to a particular animal. For example, the distributed neural networks noted by Alun Anderson (1988) in her *Nature* review were employed to simulate one of the functional maps that have been found in the visual cortex of cats. Primates have about a dozen different cortical maps of the visual world and maybe half a dozen representations of auditory events.

Recent studies of how various animals store information indicate that the brain's dictionary may really be a set of dictionaries—cortical maps—that store different aspects of real-world knowledge. Rosaleen McCarthy and E. K. Warrington (1988), for example, report the case of a patient who could neither produce or comprehend the spoken names of animate things. He had no difficulty with the names of inanimate objects or with the names of animate things when he described pictures. This modality- and category-specific deficit in accessing his mental dictionary apparently derived from an abnormality in the left hemisphere of his brain. Other victims of left-hemisphere damage show impairment in specific attributes of knowledge; for example, a person may know that a canary is a bird but not that it is yellow. It is possible that the semantic categories that are implicit in words correspond to a set of functional cortical maps that form the brain's dictionary. Neurophysiology rather than philosophy thus may ultimately hold the key to understanding the semantic structure of human language.

Individual Differences

Recent research on the neurophysiology of the brain and distributed neural networks shows that no two individuals have the same brain. Electrophysiologic studies on living brains show that cortical maps are plastic; they change as the animal interacts with the world. Experiments with adult owl monkeys, for example, show that the representation of a fingertip in the cortex expands when the monkey's fingertip is intensely stimulated. The representation of the fingertip extends one millimeter into the areas of the cortex that are usually used to map the rest of the finger and part of the hand (Merzenich, 1987b). Gerald Edelman (1987) and his associates have demonstrated that genetically identical brains form different connections as they are exposed to different experiences. Edelman's computer simulations of distributed neural networks simulate the observed processes by which the microstructure of the brain responds to stimulation. Connections that are stimulated become stronger; those that receive little or no use wither away. Edelman uses the term *neural Darwinism* to describe the process by which selection for use occurs, setting up distributed neural networks that reflect life experiences as well as genetic endowment.

Experiments in Nature

Data from various experiments in nature indicate some of the relationships that may hold between the brain's dictionary and other components of human language and thought.

Mental Retardation

Despite the oft-repeated claim that *all* human beings acquire language regardless of intelligence, this is not the case. In the limiting condition, language is absent in severely retarded institutionalized adults (Wills, 1973). The individuals who lacked language also lacked basic motor skills; less retarded caretakers tied their shoes and dressed them. Down's syndrome also results in profound motor control, syntactic, and cognitive deficits (Hopmann, in preparation). Further research on the possible connection between motor skills and language in retarded people is warranted in light of the possible neural and evolutionary connection.

Alzheimer's Disease

Although the neurological bases of Alzheimer's disease are not yet clear, it appears to involve progressive, diffuse, and bilateral damage to the brain. Formal tests show that people with even mild Alzheimer's have difficulty in producing lists of words or understanding words or the morphemes that convey meaning (Bayles and Boone, 1982; Bayles and Tomoeda, 1983). Their spontaneous speech is semantically empty (Nicholas, Obler, and Helms-Estabrook, 1985). The problem seems to rest in their ability to access the "meanings" of the words. Daniel Kempler (1988) asked eight Alzheimer's subjects with varying degrees of dementia to pantomime "how to use" or "what to do with" a set of twenty objects. Both naming and pantomiming abilities were impaired in proportion to the severity of dementia. The Alzheimer's subjects were unable to ascribe meaning to words or to specific objects either in or outside of the domain of language. Kempler concludes that the lexicon is intrinsically tied to "non-linguistic cognitive functions" (1988, p. 156).

In marked contrast, syntax and speech are preserved in even moderately severe cases of Alzheimer's. Daniel Kempler, Susan Curtiss, and Catherine Jackson (1987) studied twenty carefully diagnosed Alzheimer's patients, eliciting both spontaneous speech and writing to dictation. All the subjects were monolingual speakers of English and were compared with age-matched controls. In spontaneous utterances and writing, no significant differences could be noted in the occurrence of syntactically complex sentences. The written samples involved transcribing one member of a homophone pair with either a semantic or syntactic cue that would identify the intended word. For example, the homophone pair *see/sea* was presented with semantic cues such as *look-see* and *lake-sea* and with syntactic cues such as *I see* and *the sea*. Analysis of the samples showed that the Alzheimer's patients made seven times as many errors on the basis of semantic errors as the controls, writing *lake* instead of *sea*, *hour* in place of *day*, *see* in place of *look*, and so forth.

William's Syndrome

William's syndrome (idiopathic infantile hypercalcaemia) presents a similar limiting condition. The victims of William's syndrome appear to have a linguistic deficit that is limited to the brain's dictionary. The syndrome became known to linguists when the mother of an afflicted child wrote to Noam Chomsky (who essentially equates language with syntax) and noted that her child had almost "perfect" language but was severely retarded.[2] If human language consisted of only speech and syntax, then the victims of this pathology could be said to have "perfect language."

Children with William's syndrome have a distinctive "elfin" face and a number of physical abnormalities. Their speech production is fluent, and their syntactic abilities are, if anything, precocious. They typically are overfriendly toward strangers and converse readily, but the content of their speech is empty. Parents and teachers characterize their speech as "fluent and articulate . . . but at a superficial level . . . including the use of stock phrases and intonations picked up from adults . . . good rote recall of words, phrases, stories, and tunes" (Udwin, Yule, and Martin, 1986, p. 306). William's syndrome children are clearly mentally handicapped; those studied by Udwin and his colleagues had IQs that averaged about 50 over a range of standard tests. Their verbal comprehension was extremely poor; their spatial orientation and motor control, apart from speech, were also poor.

The correlation between good speech motor control and syntax that characterizes William's syndrome children is consistent with human syntactic ability's involving brain mechanisms that initially were adapted to facilitate speech rather than general motor control. The empty content of their speech suggests that they are unable to form or access the rich families of concepts that are the cognitive reflections of the words that constitute the brain's dictionary. However, the precise nature of the linguistic and cognitive deficits and the biological basis of William's syndrome await further research.

Wernicke's Aphasia

Data derived from the study of Wernicke's aphasia are also consistent with the premise that human language involves different components. Carl Wernicke (1874), in his original observations of two patients, noted that their speech production was articulate and their syntax correct, but that they used inappropriate words or neologisms. Wernicke's patients have some problems with the phonetic addresses, that is, the sounds that signify words; they also confuse semantically related words. These problems are apparent in their use of words that are phonetically or semantically related to the words that they should have spoken. For example, a Wernicke's patient might substitute the word *light* for *lock* or *key* for *lock,* or produce the neologism *glip.* The deficit appears to be one of dictionary access. Noting the similarities in the lexical-semantic deficits that occur in Wernicke's aphasia and Alzheimer's disease, Kempler and his associates (1987) suggest this as a neurological-linguistic problem common to both.

The errors produced by Wernicke's aphasics correspond to those produced by a slightly damaged distributed neural network. Distributed neural networks when they are used to implement dictionaries can access words either by means of their phonetic addresses or through semantic association. The neural network's dictionary-search strategy conforms to what people do as they search for an appropriate word. How many times have you searched for the sound of a word whose meaning you know? That happens very often to me; I can dredge up all sorts of semantically related words, but not the right word; then finally the appropriate word pops into consciousness. Wernicke's patients seem to have a pervasive problem in this regard. Naming difficulties are one of the most pervasive linguistic deficits of all instances of aphasia. Broca's aphasics also frequently have difficulty naming objects and accessing words, but the general picture noted earlier holds, and they have more difficulty with words that have grammatical functions (Goldstein, 1948; Stuss and Benson, 1986; Caplan, 1987).

The Linguistic Dictionary and General Intelligence

Word knowledge and general intelligence may be linked at a basic biological level. Psychologists have been arguing since the beginning of this century about whether general intelligence exists and, if it exists, how it may be related to more specialized aspects of intelligence. Edward L. Thorndike (1913), for example, devised tests of various types of mental operations, then related general intelligence to the excellence or speed with which an individual completed the tests. David Wechsler (1944) developed these procedures further; his tests for the measurement of intelligence are routinely used today. Although individuals show different skills or abilities, most theories concerning the nature and development of human intelligence conclude that a *general* intellectual factor exists that affects the outcome of any *specific* skill. Carl Spearman (1904) first explicitly proposed that a general, or g, factor of intelligence operates in conjunction with task-specific skills, and that it could be calculated from a battery of tests of specific skills, which would then specify the general intelligence of an individual. This theory is explicit in the Wechsler Test of Adult Intelligence and its various modifications for children (Wechsler, 1944, pp. 1–12), as well as in recent theories that attempt to account for the nature of human intelligence (Gardner, 1983; Sternberg, 1985).

General intelligence, Spearman's g, must relate to some brain mechanism, mechanisms, or a general property of a person's brain. People learn the meanings of new words through at least two processes: association and deduction. Young children clearly do not run off to consult a dictionary every time they encounter a new word—for one thing, they cannot read at first. Nor do they continually ask their caretakers for formal definitions of new words. Children listen and observe; they make use of the real-world context and of words and syntactic rules that they have previously acquired to learn the meanings of new words (Landau and Gleitman, 1985). But it is clear that some children learn more words than others. Robert Sternberg's (1985) work offers an insight. He compared students' scores on standardized intelligence tests with the size of their vocabulary and found that the number of words that they knew, irrespective of what the words were, correlated with the test score. Brighter people learn more words. They may perhaps have better or larger distributed neural networks—if neural networks turn out to be an appropriate model for the brain's computational machinery.

On the Education of Mollusks

Although computer-implemented neural networks that simulate Hebbian synaptic modification can learn by association, it is necessary to demonstrate similar phenomena in living animals. Recent students using the marine mollusk *Aplysia* have confirmed the details of Donald Hebb's (1949) theory. These simple animals have synapses that can be readily monitored and can form associations and recall them. *Aplysia* respond to threatening situations by "running" away and by withdrawing a siphon from which they can squirt ink to befuddle likely pursuers. *Aplysia* can be trained to associate a normally pleasing stimulus, essence of shrimp, with something nasty, electric shock (Carew, Walters, and Kandel, 1981; Walters, Carew, and Kandel, 1981). The procedure used was the classic Pavlovian method. Shrimp extract was first applied to the mollusk's head for ninety seconds; an unpleasant electric shock was applied to the mollusk six seconds after it received the shrimp extract. Twenty mollusks each received six to nine of these paired stimuli—shrimp extract followed by electric shock. Twenty other mollusks were in a control group that received either unpaired shocks or shrimp extract at ninety-minute intervals.

All the mollusks were tested eighteen hours later to see what they had learned. Shrimp extract was again applied to the heads of all the *Aplysia*. Those trained with paired stimuli forcefully reacted when a very mild electric shock was applied to their tails: they "ran" away, withdrew their siphons, and ejected ink. Electrophysiologic data showed that the conditioning stimulus—the shrimp extract—enhanced the synaptic transmission of the electrical signals that triggered these motor responses. The trained *Aplysia* apparently remembered their unpleasant experiences of the day before. The control-group mollusks did not react very forcefully. Since *Aplysia* have no structure that even vaguely resembles a cortex, it is evident that Pavlov was wrong when he claimed that associative learning is exclusively cortical.

Although they do not have a cortex, or even the brain of a therapsid, *Aplysia* can remember the circumstances that are associated with unpleasant experiences. As Hebb (1949) predicted, memory is chemically coded in the synapses (Greenberg et al., 1987). The animal is first exposed to a brief series of electrical shocks applied to its tail, which results in its withdrawing its siphon more forcefully in response to other threats. These learned responses are apparent a day later and result in chemical changes in the synapses that transmit sensory information to the motor control neurons of the animal's siphon. Proteins are synthesized that facilitate the transmission of the sensor electrical signals across the synapse. In *Aplysia* memory and associative learning are linked phenomena that involve modifying the synapses.

Humans Reason, Animals Respond

There is a long tradition that uses the simple expedient of assigning different words to what animals and humans do, thus denying continuity between human and animal thought. Whereas Pavlov's dogs acquired a *conditioned reflex* when they learned to associate the sound of a bell with food, human diners form a *logical inference* when they read the words *filet mignon* on a menu (and perhaps salivate ever so slightly). The same association between an arbitrary symbol—the bell's ringing for the dogs, the words for a human—and food has a different label and so becomes a qualitative distinction between *Homo sapiens* and any other animal. Anyone who lives with a dog can see that the dog thinks, if thinking includes the ability to act on learned inferences. The rather timid large dog who lived in my home for many years would avoid the garden because he was once stung by a bee. His behavior was a classic case of one-trial learning. The bee sting occurred when he was six years old, but he remembered it until his death years later and could not be coaxed into the garden for more than a brief sortie.

I am not claiming that my dog thought as well as I do, but he clearly did think. The mollusk experiments show that they also think, insofar as some of the neural mechanisms involved in associative learning are present in these simple animals. Either we must accept the proposition that they "think" to some degree, or we must arbitrarily decide that associative learning is not a cognitive act.

Summary

The experiments, observations, and computer simulations discussed in this chapter indicate that the dictionary or dictionaries of the brain are independent from the neural mechanisms that regulate syntax and speech production. These data also indicate that the brain mechanisms that underlie human linguistic ability play a part in other aspects of cognitive behavior. The semantic aspect of the brain's lexicon is in a sense our storehouse of real-world knowledge, which we acquire by means of the brain mechanism that constitutes this dictionary. There appears to be no sharp dichotomy between animals and humans in this regard. The distributed neural networks that are the best present guess concerning the neuroanatomy that makes this associative dictionary possible, are present to some degree in virtually all animals. Although the distributed neural networks that we have been discussing are only crude approximations to the neuronal networks of life, they provide a starting point for understanding the way that humans learn the most unique aspect of their behavior—language.

5

Learning to Talk and Think

The acquisition of a child's first language is one of the mysteries of human life. Although children need years of schooling to learn arithmetic, geometry, or history, they effortlessly acquire the ability to use a system of communication that learned scholars cannot adequately describe. Children, moreover, acquire their first language in virtually any environment. How do children learn language? Do they use the general cognitive strategies that they apply to other problems, or is there anything special about the process? It is apparent that evolution of human language involves an admixture of biological mechanisms specific to *Homo sapiens* and language, with ones present in other intelligent animals. I will argue for a similar basis to the process by which a child acquires language. Humans clearly have innate mechanisms that facilitate the acquisition of language, but children also use the cognitive strategies and processes that evolved with mammals—play, imitation—as well as phylogenetically older processes such as associative learning.

Universal Grammar

The difficulties that attend the acquisition of language are apparent when one considers the production and perception of speech, but the present focus of linguistic research is syntax. There is nothing very mysterious about syntax. As the tutor in Molière's *Le bourgeois gentilhomme* said to his pupil, "You already know everything there is to know about syntax." In most languages, the words of a sentence must appear in a certain order and with specific properties. English, for example, has many syntactic constraints. It is an SVO language: the subject (S) usually precedes the verb (V) and object (O). The verb and subject of an English sentence must "agree": if the subject is plural, the verb must be, too. Other, more complex restrictions also exist. The syntactic rules devised by linguists codify these relations, which speakers of English already "unconsciously know." However, it is very difficult to describe the linguistic facts; no one has yet been able to describe adequately all the syntactic rules of even one language. Linguists have also convinced themselves that young children struggling to learn their first language hear a garbled mixture of syntactically correct and wrong utterances—the sort of speech that learned scholars use. So the problem faced by a young child would seem almost insurmountable: the input is disordered; the rules are so complex that no child, however precocious and bright, could learn them. Therefore, many linguists claim that children really do not learn the rules of syntax; they must be innate.

According to Noam Chomsky, the principles that allow a child to acquire syntax are built into the child's brain in the form of a hypothetical *Universal Grammar*,[1] which "may be regarded as a characterization of the genetically determined language faculty. One may think of this faculty as a 'language acquisition device,' an innate component of the human mind that yields a particular language through interaction with presented experience" (1986, p. 3). Moreover, the Universal Grammar is independent from other aspects of human cognitive ability:

> there is little hope in accounting for our [linguistic] knowledge in terms of such ideas as analogy, induction, association, reliable procedures, good reasons, and justification . . . or in terms of "generalized learning mechanisms" . . . We should, so it appears, think of knowledge of language as a certain state of the mind/brain . . . some distinguishable faculty of the mind—the language faculty—with its specific properties, structure, and organization, one "module" of the mind. (1986, p. 12)

According to Chomsky, language "must be largely or completely deductible from general principles, because relevant information is unavailable to the language learner" (1986, p. 105).[2]

Many recent linguistic analyses of the way that children acquire the syntax of their native language accept the Universal Grammar as a given. The Universal Grammar, UG, consists of tightly interlocked operations that are characterized by notation derived from formal mathematical logic. Chomsky (1986, pp. 145–273) characterizes it as a set of "principles and parameters." The "principles" include such elements as "a binding theory, a theta theory, a case theory," and so forth.

A Universal Grammar for Pedestrians

Despite the complex terminology used to describe the properties of the Universal Grammar, the theory is not very different in principle from many that regulate everyday life. The relation between the hypothetical Universal Grammar and a specific language is similar in many ways to the one between taking a particular walk in the city and one's prior knowledge of the street plan and traffic regulations.

Suppose that you are in New York City, standing with a friend on the southeast corner of Third Avenue and 57th Street, and want to go to the Whitney Museum, at Lexington Avenue and 72nd Street. You can walk either north toward 72nd Street or west toward Lexington Avenue. However, the traffic light is green in the north-south direction, so you both cross 57th Street. Your friend wants to look at the window displays of the chic stores and art galleries on 57th Street, so you both turn left and walk along 57th Street. Since time is running short, you turn right at the corner of Lexington Avenue and walk directly to the museum. On another walk, on a different day, with a different friend, who wanted to look at the menus of ethnic restaurants, you might take a different route: starting from the same point, you would deliberately walk up Third Avenue to 72nd Street, then turn left and walk directly to the museum. If you were instead alone and short of time you might take a zigzag course, keeping to the avenues (on which you thought you could walk faster) and turning onto a crosstown street when the avenue traffic light was red. Your particular route is det∍rmined by two factors: the specific circumstances of the day—your friend's interests and the traffic light's turning green at a particular moment—plus prior knowledge of the con-straints of the street plan and traffic principles. You cannot tunnel through buildings or run heedlessly against the traffic. The general principles that constrain all walks interact with par-ticular events to yield a particular walk.

Similarly, in the theory of the Universal Grammar, Chom-sky's "markedness system" automatically selects certain sets of principles contingent on an initial decision. For example, if you decided that time was very short and hailed a taxicab, the markedness system would bring into play the one-way-street constraints that apply to vehicles. You could not drive uptown from 57th Street to 72nd Street on Fifth Avenue even if you wanted to look at the soothing greenery of Central Park—the route is one-way in the opposite direction.

Why Current Universal Grammar Is Biologically Implausible

Many of the data discussed in the previous chapters are consis-tent with humans' having innate brain mechanisms that facili-tate and structure the acquisition of language. The idea that an innate Universal Grammar determines some aspects of syntax is therefore, in itself, a reasonable hypothesis and presents a goal for research that would determine precisely what aspects of syntax might have an innate, genetically transmitted basis. However, any innate Universal Grammar must not violate the general constraints that determine the properties of all geneti-cally transmitted biological organs.

The current version of Universal Grammar clearly does not meet this criterion. It assumes that all human beings are born with an *identical* "plan" that contains all the principles and constraints on which any human language is constructed. Fur-thermore, it consists of a tightly interlocked set of "principles," "components," "conventions," and so on. Every one of these must be present in the individual child's brain for the UG to work. A child could not learn a particular language unless the *complete* system was in place. This would lead to rather peculiar results if children really did acquire their first language by means of the Universal Grammar. Although nativist linguistic theory is clothed in the garb of biology, it overlooks a key principle of biology that has been understood since Darwin's time—genetic variation.

Genetic variation is the feedstock of evolution. No two individuals save identical twins, triplets, and so on are similar. Darwin took note of the variation that animal breeders and farmers have observed and used for thousands of years. If a farmer, for example, wants to develop a variety of corn that is more resistant to drought conditions, he *selects* the corn plants that thrive with less water and plants their progeny. In the population of corn plants there is always genetic variation that expresses itself in terms of resistance to drought. The same principles hold for animals; if a farmer wants to breed cows that deliver more milk he selects the progeny of the cows who yield more milk under the particular environmental conditions that are relevant. Contrary to some popularized accounts of evolution, the farmer does not have to wait for a mutation or try to foster one; genetic variation is always present. Modern molecular biology confirms Darwin's observations concerning the pervasive variation that characterizes living organisms: the genes at the chromosomal location that determines some particular aspect of the structure of a mammal vary about 10 percent of the time (Mayr, 1982).

Suppose that nativist theory were correct and all children had a detailed, genetically coded "language faculty" that was absolutely necessary to acquire any language. Then it would follow that some children would lack one or more of the genetically coded components of the language faculty. Some "general principle" or some component of the "markedness system" *would necessarily be absent in some children because it is genetically transmitted.* This is the case for all genetically coded aspects of the morphology of human beings or any other living organisms. Color-blind or color-deficient people, for example, manifest one aspect of genetic variation. Color blindness results because color vision is, in fact, determined by innate, genetically transmitted color receptors. A person who lacks the appropriate genetically transmitted color receptors will be color blind. If the acquisition of language were as tightly constrained by the genetic code as color vision, then we would find similar anomalies.

Chomsky in fact unintentionally points out a test of his own theory when he notes that "passivization in English (but not some other languages, such as German) is generally limited to transitive verbs; hence, *ask* but not *wonder* or *care*" (1986, p. 88). For example, the passive sentence *John was hit by Susan* is grammatically correct, whereas *John was wondered by Susan* is not. The acquisition of passive sentences by children learning English supposedly does not follow from their taking note of various examples of passive sentences that they may hear and then deducing the grammatical "rule." According to Chomsky, their acquisition of the correct form of English passives derives instead from the genetically transmitted "case" principles that are invoked in English. If this were really so, then children who lacked these particular principles, even though they were raised in an English-speaking environment, would not be able to learn to form correct passive sentences. They theoretically would be able to easily learn German passives, which do not invoke this missing component of the genetically transmitted "case theory" (Chomsky, 1986, pp. 84–105). An experiment that would support Chomsky's Universal Grammar therefore would first find a group of children raised in an English-speaking environment who were unable to form grammatical passives. If the children were then exposed to German and readily acquired correct German passives we would have indirect proof of a genetically transmitted case principle. If the case principle had a strong genetic component, some of the children's mothers or fathers should also show these effects.

A biologically plausible Universal Grammar cannot have rules and parameters that are so tightly interlocked that the absence of any single bit of putative innate knowledge makes it impossible for the child to acquire a particular language. In other words, we cannot claim that a single set of innate principles concerning language exists that is (a) absolutely necessary for the acquisition of language and (b) uniform for all human beings. The usual defense offered by linguists for claim b—genetic uniformity—is that "all people have a brain, a heart, a nose, two arms, two legs, and two eyes" and so forth. But it is as plain as the nose on your face that all noses vary, all eyes vary, and that that is also the case for hearts, lungs, and brains. Although some aspects of language might be highly "buffered," transmitted by multiple genes in a manner that minimizes the effects of variation, the problem is to determine their nature. It does not further research to ignore biological variation and instead to formulate a Universal Grammar consisting of a tightly nested set of principles and procedures all of which must exist.[3]

The biological implausibility of the current nativist linguistic theory may be more readily appreciated if we were to propose a similar theory for learning to drive a car. A nativist theory similar to the currently hypothesized Universal Grammar would claim that a "Universal Driving Capacity" exists that consists of a tightly interlocked set of "principles," "components," "conventions," and so on. The Universal Driving Capacity would contain genetically coded principles that could account for all possible driving styles, including a stop-on-red principle, a right-turn-on-red principle, a four-way-stop principle, and a maximum-speed principle. Selectional principles and conventions would ensure that a learner-driver selected a correct set of "rules" that would yield appropriate maneuvers in the particular driving style observed. Although different driving styles exist, a learner-driver hypothetically would be able to acquire a particular driving style by observing the behavior of adult drivers, thereby activating various elements in the genetically coded Universal Driving Capacity. But suppose that some element of the genetically coded Universal Driving Capacity—say, the stop-on-red principle—was not present in an individual learner-driver. Then that unfortunate learner-driver would never be allowed to drive. A learner-driver who lacked the right-turn-on-red principle would likewise be unable to acquire a driving style that invoked this principle. However, this genetic deficit would not be as serious: although the learner-driver would never be able to drive in California, he or she would be able to drive in Boston.

How Children Might Learn Language

The Input

Although studies of the speech that is directed to young children date back to the 1930s, only in the last decade or so has it become apparent that some children do hear a great deal of grammatically correct speech. For example, Elisa Newport, Lila Gleitman, and Henry Gleitman (1977) found that about 90 percent of the speech directed to young children by the college-educated middle-class American mothers they studied was grammatical. Catherine Snow (1977), who studied "motherese" in detail, noted similar results. Parents make subtle adjustments in their speech when they talk to children, often at levels that can be discerned only through acoustic instrumental analysis. Ann Fernald (1982), for example, showed that mothers alter the pitch of their voice when they speak to their newborns; the fundamental frequency of phonation is higher and sweeps through a range of almost two octaves. The exaggerated intonation serves as a "directing" signal that highlights the speech addressed to the child. Many mothers continue to do this until the child is two or three years old. A similar pattern exists among speakers of Chinese: in this case the mother also has to produce the correct "tones"—fundamental frequency patterns that specify different words; the phenomenon may well be a human "universal" (Grieser and Kuhl, 1988).

However, we still lack comprehensive data on motherese and virtually all other aspects of language acquisition by children raised in different cultural settings. For example, the parents of children raised in Micronesia hardly ever talk to them during the first few years of life (Schieffelin, 1982). Differences in parent-child interaction exist even in middle-class American society. Snow (1977), for example, found that the children she studied heard extremely simplified sentences from their parents;

she concludes that this simplified input may have facilitated their learning the rudiments of syntax. Other studies found a greater variety of sentences and concluded that the syntactic complexity of the speech directed to children has very little effect on their language acquisition (Newport, Gleitman, and Gleitman, 1977). Many other factors that vary from child to child appear to affect the acquisition of language. Some studies claim that language acquisition depends on engaging in "conversations" that involve the joint attention of mother and child: some common specified object or activity must be discussed (Tomasello and Farrar, 1986). For example, both must be looking at a dog while the mother says *Look at the dog*—a clear example of associative learning.

It is also clear that mothers vary in the amount of attention they give their children, and it is possible that extra attention and "better" language input provide at least an initial advantage in the acquisition of language. Michael Tomasello and his colleagues, for example, have shown that mothers cannot provide as much joint attention to one-to-two-year-old twins as to single children. Perhaps as a result the twins were lower than single children on all measures of linguistic development at age twenty-one months (Tomasello, Mannle, and Kruger, 1986). However, it is doubtful that twins suffer any lasting linguistic impairment. Psychologists such as Jerome Kagan and Melissa Bowerman, who have spent decades studying children growing up in different cultures, do not find any obvious differences in the rate at which children acquire language or in their ultimate proficiency. Melissa Bowerman (1987, 1988) has studied children acquiring many different languages, including German, English, Dutch, Spanish, and Korean. In a 1987 review article she concludes that children can use virtually all the cognitive strategies that have been proposed in the last twenty years of intense study of language learning. Different strategies apparently are used by different children in different settings, but they all seem to work. Language acquisition seems to be a highly buffered mechanism that occurs under most normal circumstances, that is, in situations involving social interaction using language.[4]

Concept Formation

One such cognitive strategy is *concept formation,* or rule learning. Children do not need to be presented with a "perfect" input in which all the utterances that they hear are grammatically correct. Like the distributed neural networks discussed earlier, they simply need an input in which most examples are correct. From these examples they can generalize to derive the syntactic rules that underlie a set of utterances that show a regular pattern, correcting for occasional errors. Jenny Singleton and Elisa Newport (1989) studied a deaf child acquiring his first language, American Sign Language (ASL), from poor input. Deaf children sometimes acquire language solely from exposure to their parents; the parents of the child who was studied, though deaf, were not proficient in ASL. ASL has a very different syntax from that of English. Its verbs are particularly complex and are marked for distinctions that are not conveyed in English verbs. Various complex hand gestures serve as morphemes that signal continued versus interrupted activity, the shape of an object that is being acted on, and so forth.

The parents made errors about 40 percent of the time when they formed various ASL verbs. The child was much better than his parents; he made about the same number of errors (20 percent) on these verbs as deaf children learning ASL from proficient adult ASL users, who are almost error-free. He generally used the correct regular morphemes that modify a verb even though his parents often used an inappropriate one. An equivalent example for speakers of English would be a parent who sometimes correctly added *ed* to *look* to form *looked,* but at other times said *looking* when he meant to say *looked.* An equivalent hypothetical English-speaking child in this situation would learn to say *looked* consistently.

The child's behavior seemingly supports the claim of nativist linguistic theory that the role of the speech input is merely to "activate" the innate Universal Grammar (Pinker, 1984; Chomsky, 1986); the UG overrides the poor input that the child hears. The deaf child presumably does not pay attention to the parents' ill-formed ASL because the principles stored in the Universal Grammar guide him to the correct set of grammatical rules. However, Singleton and Newport found nothing that resembled the activation of a Universal Grammar. The child instead appeared to be monitoring his parents' ASL verbs, consistently adopting the morpheme that his parents used *most often*. Since his parents were usually correct 60 percent of the time, the child generalized the morpheme that they used most frequently to acquire a language that was more correct than his parents'. The crucial data concerned the cases in which his parents never used the correct morpheme more than 50 percent of the time; the child then acquired the incorrect morpheme or simply omitted it. Clearly, the child used a cognitive strategy, forming an "archetype" or generalization on the expectation that language has a logical structure. Recent data on the acquisition of vowel sounds by infants also support the relevance of general cognitive strategies: six-month-old infants identify vowels better when they can compare sounds with well-formed prototypes (Grieser and Kuhl, 1989).

Critical Periods

One aspect of language acquisition that clearly has an innate biological basis is the *critical period*, that is, an age after which it is almost impossible to attain native levels of proficiency. Virtually everyone studying language has agreed that a critical period exists (for example, Lenneberg, 1967), but until recently no one was able to demonstrate this in a scientific manner, because most tests would involve deliberately isolating a child from normal language beyond the critical period—which would constitute unethical behavior. However, Elisa Newport and her husband, Ted Suppala (1987), have demonstrated that a critical period does exist for language learning. Newport and Suppala studied an experiment in nature, deaf children born to hearing parents and exposed to ASL at different ages. The comprehension of various aspects of language of fifty-to-seventy-year-old deaf people was tested; their error rates depended on when they had first learned ASL. "Native" users of ASL, who had been exposed to ASL from birth onward, had extremely low (2–5 percent) error rates for various aspects of the grammar and word structure. There was only a small increase in the error rate for people who had acquired ASL between ages four and seven. The error rate was 24 percent higher for those who had learned ASL as adolescents. Ted Suppala is deaf, so their own hearing children are learning ASL as well as English from birth onward.

Similar effects were observed for second-language acquisition. Jacqueline Johnson and Elisa Newport (1989) compared the English proficiency of forty-six native speakers of Korean or Chinese who had arrived in the United States between the ages of three and thirty-nine and had lived in the United States between three and twenty-six years at the time of testing. The subjects were asked to state whether various English sentences were or were not grammatical; a wide variety of structures of English grammar was presented. People who had arrived in the United States before age seven did best; their performance was equal to that of native-born speakers of English. Test performance fell gradually with age of arrival for subjects who had arrived between age seven and puberty. The subjects who came to the United States after age seventeen averaged about 22 percent more errors than those who had come before age seven. The subjects were also interviewed to assess their motivation, amount of experience with English, and American identification. None of these affected the age effect.

Associative Learning

Critical periods are not limited to language acquisition. Hundreds of independent neurophysiological experiments have shown that critical periods limit the acquisition of the perception of visual form in cats. The effects of age have long been noted by animal trainers. The old saw "You can't teach an old dog new tricks" does, after all, hold for dogs as well as for people. The process of learning language probably involves the sort of rewardless associative learning that underlies the way dogs and other animals acquire knowledge of the world.

There is a general misconception that associative learning is equivalent to the concept of reinforcement introduced by the noted behavioral psychologist B. F. Skinner. There is also a misconception that Chomsky's 1959 critique of Skinner's behavioral theories proved that associative learning could not account for the way that children acquire the principles that underlie syntax. According to Skinner (1953), animals learn by means of either positive reinforcement or negative reinforcement. If, for example, you give your dog a treat when he sits down on verbal command, you are using positive reinforcement. If instead you wired your dog to an electric shock machine, said *sit*, and shocked him when he did not sit, you would be using negative reinforcement. Skinner elaborated a theory of operant conditioning that claimed that certain schedules of reinforcement—certain intervals between trials and rewards—were better than others. Skinner claimed that almost all aspects of animal and human behavior, within the limits imposed by anatomy and physiology, could be modified by this technique.

Many ingenious tricks were taught to animals by means of these techniques, guiding them step by step. Each step involves rewarding the animal by feeding it when it performs the task. A bravura performance by a trained rat involved the rat's going through a doorway, climbing a spiral staircase, pushing down a drawbridge, crossing a bridge, climbing a ladder, pedaling a model railroad car over a track, climbing a flight of stairs, playing on a toy piano, running through a tunnel, stepping into an elevator, and pulling on a chain to lower the elevator to a starting point, where the rat finally pushed a lever to receive the reward—a food pellet (Bachrach and Karen, 1969). Does this mean that the rat performed all these steps because of the reinforcing pellets of food? Allen and Beatrix Gardner (1988) argue that the training instead follows from the simple contingency of the food and the activity. Gardner and Gardner (1988) apply complex formal logic to support their claim, but the gist of their argument is that simple association accounts for the rat's learning to do all these different peculiar things.

In other words, simple association would account for the way in which rats learn to perform peculiar tasks, as well as many aspects of language acquisition by children. Associative learning is not limited to immediate action; it can also be used to acquire underlying principles, whether implemented by distributed neural networks (Rumelhart et al., 1986; Sejnowski, Koch, and Churchland, 1988) or the brains of animals (Gardner and Gardner, 1988). Chomsky's (1959) review of Skinner's theories is often taken as demonstrating that associative learning cannot account for the way that children learn the concepts that underlie behavior. However, that is not the case—associative learning can reveal the principles, the "rules" that underlie specific instances of behavior.

Play, Imitation, and Curiosity

Owing to the influence of Skinner's theories, linguists and psychologists studying the role of learning on the acquisition of language often focus on the presence or absence of positive rewards or negative reprimands. Negative information is that which tells you when you're wrong. For example, if someone told you that the sentence *I seed John* was incorrect, it might facilitate your learning the correct form of the past tense of the verb *see*. Children do not appear to be overtly corrected when they begin to talk, which leads some linguists to conclude that general cognitive strategies are irrelevant. However, children have other ways of monitoring their behavior while they play at language.

One of the ways in which mammals differ from reptiles is that they play (MacLean, 1967, 1973, 1985, 1986). Another way in which they differ is that mammals are more adaptable and learn more. Play seems to be an important mechanism for facilitating learning (Baldwin and Baldwin, 1977; Lieberman, 1984), and what do children do when they play? They imitate what they observe. Imitation is another aspect of general mammalian behavior that has an adaptive value. Imitation provides implicit negative information—by observation you can compare your behavior with that of the "others" and do what they do. Andrew Meltzoff (1988) shows that humans, not cats, deserve the appellative "copy cat." Other species may be more proficient than people in some specific behavior such as imitating bird songs (Marler and Tamura, 1964). Many other species are also capable of imitating social behavior (Zentall and Galef, 1987). However, Andrew Meltzoff (1988) argues that people are clearly the most gifted of imitators among animals.

Humans are literally born imitators. Newborns, for example, will imitate adult facial expressions (Meltzoff and Moore, 1977, 1983). The photographs in Figure 5–1 show Andrew Meltzoff making faces at newborn infants and the infants imitating him in three different experimental sessions. The newborn can see only Meltzoff, who is positioned in front of a blank white screen to block out distracting things. Meltzoff then sticks out his tongue and makes other faces at each infant. Simultaneous video recordings of Meltzoff and each infant show that the infant is imitating Meltzoff, not the other way round. These experiments show that by age nine months, infants will imitate activities that they saw a day before, such as pressing a button on a box that sounds a buzzer, or shaking an egg-shaped object that makes a noise. By age fourteen months infants will imitate

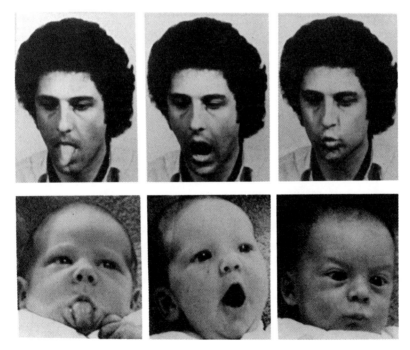

Figure 5–1. Sample photographs from videotaped recordings showing Andrew Meltzoff making faces at infants and the infants imitating tongue protrusion, mouth opening, and lip protrusion. (After Meltzoff and Moore, 1977)

a completely novel act that they saw a week ago. Why do they bother to do this? It's play!

Imitation has long been recognized to have an adaptive value for humans in acquiring the details of complex culture (Washburn, 1961; Piaget, 1962; Huxley, 1963; Lorenz, 1974; Bruner, 1983). Imitation contributes to biological fitness in the animals who are most closely related to humans. Chimpanzees, whose use of tools is closest to humans', learn by imitating other chimpanzees (Beck, 1974; Goodall, 1986). Jane Goodall at the Gombe Stream Reserve (now Gombe National Park) in Tanzania observed and photographed a young female chimpanzee learning to use a stick to fish for termites by watching her mother. Controlled experiments with captive chimpanzees show that observation and imitation also play a part in their acquiring new patterns of tool use. Chimpanzees at the Yerkes Center in Georgia learned to use a simple T-shaped stick to pull food into their cage when they saw an adult use the tool (Tomasello et al., 1987).

Imitation is probably the most important mechanism for the transmission of human culture. There is no need to postulate a special-purpose innate fork-using brain mechanism to account for the way that children learn to use forks, or universal clothes or car grammars to account for the way that people rush to outfit themselves in the latest style in clothes or cars. Imitation and a desire to "be like the others" clearly can account for most of the short-term changes in human culture, and perhaps for many of its major achievements.

The issue we have been discussing is not whether human linguistic ability ultimately involves specialized brain mechanisms. Specialized brain circuits clearly underlie speech production and syntax. However, we probably learn many, if not all, of the automatized motor control patterns and rules of syntax by means of general cognitive mechanisms.

Language and Thought

The Sapir-Whorf Hypothesis

One of the major issues that continue to divide scholars is the relationship between language and thought. The debate derives from a proposition framed by Wilhelm von Humboldt in 1836 and reaffirmed a century later by the American linguists Edward Sapir (1949) and Benjamin Whorf (1956). According to the Sapir-Whorf theory, differences between languages are reflected in the way the native speakers of these languages think and reason about the world. The Zuni language, for example, lacks words that differentiate the colors yellow and orange. An early experiment by Roger Brown and Eric Lenneberg (1954) seemed to support this view: native Zuni speakers failed to differentiate these colors, whereas English speakers had no difficulty in performing the same task. However, subsequent research showed that the results derived from the native Zuni subjects' unfamiliarity with test taking rather than from any basic cognitive or perceptual difference.

In fact the color names used in a wide range of languages seem to derive from the biological properties of the primate visual system (Berlin and Kay, 1969). We have three basic color receptors in our visual system. The basic color names roughly correspond to the way these receptors respond to light. Languages that do not explicitly code hue use the terms *black* versus *white* or *dark* versus *light*. Some explicit color names are *red*, *blue, green*, and *yellow*. Tests of speakers of languages that lack these explicit color names show no differences in color perception from speakers of languages that explicitly code these colors (Heider, 1972).

Similar results obtain for more abstract concepts. Chinese, for example, lacks an explicit subjunctive mood, which allows a speaker to state a hypothetical or counterfactual contingency, such as *If I were king I would . . .* However, monolingual Chinese and English speakers do not differ materially in their ability to comprehend or express this distinction (Au, 1983). Again, initial data appeared to support the strong version of the Sapir-Whorf hypothesis, that cognition is strongly influenced by language. Alfred Bloom (1981) found that monolingual Chinese speakers had difficulty in comprehending subjunctive sentences in a Chinese rendition of a short English story. However, no differences between English and Chinese subjects were apparent when a more idiomatic translation of the same story was used by Terry Au, who is a bilingual speaker of Chinese and

English. The strong version of the Sapir-Whorf hypothesis thus is wrong; the ultimate limits on the cognitive ability of Chinese and English-speaking individuals seem to be similar.

What remains unclear is whether the form of one's language affects how a person views and reacts to the world at any given moment. In all the experiments discussed above, the subject necessarily focuses on a particular cognitive concept or percept. The form and vocabulary of a language inherently provide this focus. I can certainly determine the texture and softness of various forms of snow, but lacking the Eskimo words and corresponding *semantic categories* that characterize these words, am I likely to notice these different types of snow? Will people whose language does not code the subjunctive think along these lines as often as they might if the language possessed that form? These questions should be addressed in the light of many studies of language acquisition that show a strong link between language and thought.

Hierarchical Categorization

Although the form of a particular language does not necessarily set a limit on the nature of the speaker's thought, developmental studies of children indicate strong interactions between language and thought. For example, hierarchical categorization systems allow people to organize their knowledge. Success in parlor games such as Twenty Questions is based on sagacious use of hierarchical categorization: when the answer is *animal* in reply to the question *Is it animal, vegetable, or mineral?* your second question should not be *Is it a mountain?* Hierarchical categorization systems implicitly code knowledge. When someone tells you that the item on the counter is a fruit, you immediately know a great deal about it. You know that it is a plant, that it can be eaten, and that it probably tastes sweet.

This powerful cognitive mechanism is built into the structure of human language. As Sandra Waxman (1985) notes, it provides a framework for much of our logical reasoning. In a series of categorization experiments with three- and four-year-old children, Waxman showed that children know that adjectives modify nouns and furthermore know that certain morphemes indicate that a word is an adjective. Knowing how to recognize that a word is an adjective and knowing what adjectives do facilitated the children's ability to categorize sets of photographs. The children were asked to sort photographs of various animals and food. In different experiments they were prompted to sort the photographs into the superordinate categories of animals versus food, the basic level of dogs versus cats, and the subordinate level of big dogs versus little dogs. The children were better able to classify the photographs at the superordinate level when the experimenter described each photograph using a Japanese-like noun conforming to the sound pattern of English. Waxman would, for example, point to a picture and say that *This dog is a suikah.* This did not facilitate their subordinate sort into big versus little dogs. In contrast, saying *This dog is suk-ish* facilitated their sorting the photographs into the appropriate subcategories. The children were aware of the hierarchical aspect of English morphology—if something is Xish (an adjective), then that property pertains to a subcategory. The linguistic system of English, which codes subcategories as adjectives, facilitated the cognitive task.

Virtually all studies of language development show that a "naming explosion" occurs at about age eighteen months (Bloom, 1973; Nelson, 1975). Children develop an interest in naming things, and their vocabulary suddenly begins to expand. This interest in naming coincides with a cognitive milestone, the development of categorization. Children between ages fourteen and eighteen months spontaneously begin to sort different-shaped objects into piles, putting balls into one pile, boxes into another. Experiments show that the naming explosion occurs shortly after children are able to categorize objects or to infer means-ends understanding such as using a vertical string to obtain an object (Gopnik and Meltzoff, 1987). Other aspects of categorization are more abstract. For example, children begin to use the word *gone* when they acquire the concept of object-permanence. The classic test of object-permanence is a child's version of the old shell game: a ball is concealed beneath a piece of cloth or a cup. Alison Gopnik and Andrew Meltzoff (1985) have shown that children first use the word *gone* when they know that the covered ball does not simply disappear. Children "acquire" the word *gone* when they acquire the concept that the word *gone* signifies. Words reflect thoughts, and their development proceeds in concert.

Syntax, Morphology, and Cognitive Processes

Evidence is emerging that indicates that even very young children are aware of the cognitive implications of English syntax and morphology. Kathy Hirsch-Pasek et al. (1988) showed that eighteen-month-old children comprehend the distinctions conveyed by the prepositions *up, on,* and *under.* Animated cartoons synchronized with spoken messages such as *Look at the bear running on the table* were presented on television screens. One cartoon showed a bear running on a table, the other a bear going under a table. The children consistently looked more often at the appropriate cartoon. The mechanical presentation avoided the cuing effects that have invalidated so many earlier studies of early comprehension of language by children.[5]

Barbara Landau and Lila Gleitman (1985) make a strong case for the idea that young children use their knowledge of the cognitive relations conveyed by the syntax of English to help them acquire the meaning of words. Children first use a number of cues to derive the syntactic structure of language. For example, they appear to use the basic intonational breath-group signal to segment the flow of speech into phrases; stressed words are also acquired first (Gleitman et al., 1987).[6] Once they understand the rudiments of syntax, they apply this knowledge to aid in their understanding of words. For example, a child may determine that a word is a noun by noting its position in a simple phrase such as *See the X.* Blind children acquire some of the meanings of words such as *look* and *see* by this process, which Landau and Gleitman term "syntactic bootstrapping." In short, there seems to be a strong link between cognition and language; the cognitive distinctions conveyed by language are available at a very early age.

Bilingual Children

An odd twist to the language-thought debate is the discovery that children's knowledge of a second language seems to result in better performance of nonlinguistic cognitive tasks. Through the 1960s it was generally thought that, all other things being equal, bilingual children had less developed linguistic ability in either language; bilingualism was also supposed to have a general negative effect on intellectual functioning (Carringer, 1974). This view has dramatically shifted. Earlier studies often compared bilingual children with monolingual children without taking into account the subjects' age, sex, socioeconomic status, or degree of bilingualism. Many independent studies show that bilingual children perform better on tests that explore the ability to form categories, manipulate symbols, derive inferences, and follow complex instructions. These studies have included children who were bilingual speakers of a number of languages—Afrikaans and English (Ianco-Worall, 1972), Hebrew and English (Ben-Zeev, 1977), Spanish and English (Carringer, 1974; Powers and Lopez, 1985), French and English (Peal and Lambert, 1962; Bain and Yu, 1980), Welsh and English and Nigerian and English (Okoh, 1980), Kond and Hindi (Mohanty and Pattnaik, 1984), and German and English and English and Chinese (Bain and Yu, 1980). The general conclusion is that young bilingual children at first have smaller vocabularies in each language than monolingual children. This effect led earlier studies that focused on vocabulary size to conclude that learning two languages resulted in a linguistic and perhaps a cognitive deficit. However, bilingual children are consistently two to three years ahead of their monolingual contemporaries in tests that involve cognitive skills or their knowledge of the structure of language.

Anita Ianco-Worall (1972), for example, showed that four-to-six-year-old Afrikaans-English bilinguals grouped words on the basis of meaning rather than on the basis of sound pattern. Thus, they grouped *cap* and *hat* in a test in which they were asked, "Which is more like cap, can or hat?" The matched monolingual children did not do this until two or three years later. The bilinguals also did much better in tasks that involved their understanding that names are arbitrarily assigned to things. Sandra Ben-Zeev (1977) used a series of tests with children between five and a half and eight and a half. Bilingual children showed more advanced processing of verbal materials, more discriminating perceptual decisions, and more propensity to search for structure in perceptual tests. In one test the children had to describe a set of nine cylinders that differed with respect to height and diameter. Ajit Mohanty and Kabita Pattnaik (1980) showed similar results with Indian children, using similar tests. The bilingual children were also ahead of their monolingual peers on tests that measured their ability to perceive rhymes, define words, create new words, and substitute symbols.

Bilingualism does not seem to confer an absolute cognitive advantage; monolingual children catch up. However, learning a second language within the critical period seems to enhance cognitive ability. The effect is what might be expected if acquiring a language involves exercising the same brain mechanisms that enter into cognition.

Summary

General cognitive processes clearly play a major role in the way that children acquire language, and the development of linguistic and cognitive ability in human infants and children appears to be linked. Although innate, genetically transmitted brain mechanisms undoubtedly facilitate the acquisition of language, these mechanisms must conform to the constraints of biology. Language and thought, moreover, appear to reflect similar biological constraints; particular languages do not inherently constrain human thought, because both capacities appear to involve closely related brain mechanisms. Although language has different components such as words and syntax, the development of each linguistic capacity has cognitive consequences. The development of word knowledge appears to reflect real-world knowledge; syntactic structure and logical principles appear to be connected.

6

Culture and Selfless Behavior

One of the many unanswered questions concerning the brain is, precisely what makes one person more or less intelligent than another? Several theories have attempted to relate factors such as brain lateralization to specific developmental deficits such as dyslexia and stuttering (for example, Geschwind and Behan, 1984), but the verdict is still out. Although it is possible to associate certain patterns of brain damage and disease with particular behavioral deficits, no neurologist would attempt to predict the intelligence of healthy persons by examining their brains. Given the present state of our knowledge, the only measure of a person's intelligence is what that person actually does. So although detailed observations of the intact brains of fossil hominids would be extremely useful, observations of their intelligent behavior would be more to the point.

We do not have any intact fossilized brains, nor can we directly observe the intelligent behavior of fossil hominids. But even though it is incomplete, the archaeological record allows us to assess the intelligent behavior of our distant ancestors by interpreting surviving artifacts of stone and bone to deduce how people lived and worked.

In interpreting the archaeological record we have to be careful to avoid, insofar as possible, interpreting the evidence from the perspective of modern humans. For example, when we see stone tools we might assume that their makers passed on the tradition of toolmaking by both explicit verbal expositions and nonverbal demonstrations. That is not necessarily the case;

some archaic hominids may have transmitted their culture without using anything like human language. Fortunately, we can assess this possibility by observing chimpanzees, who possess and transmit a fairly elaborate culture although they lack human language and human cognitive ability. Since chimpanzees have brains that are closer in size to those of the earliest hominids we know of, they also represent a sort of limiting condition— animals that are as close as possible to the earliest hominids, without being hominids.[1] Chimpanzees are also as close genetically to human beings as any other living animal; they are closer to us than related species of rabbits are to each other (Sarich, 1974).

Chimpanzee Culture

Thanks to the dedicated work of Jane Goodall (1986) and other patient observers who followed her example, we now have some idea of chimpanzee culture. Goodall's vivid, detailed picture of the things that chimpanzees can do *without* human language provides a baseline for evaluating the archaeological evidence of human culture, particularly that of early hominids, preventing us from making false claims that the presence of behavior *x* shows the presence of human language and human cognition.

Tools and Toolmaking

Chimpanzees use and make tools. Goodall (1986) has filmed chimpanzees at the Gombe Stream Reserve in Tanzania using leaf "sponges" to soak up water and using sticks to fish up termites, one of their important food sources. Chimpanzees make these termite sticks by stripping leaves off small branches. Young chimpanzees learn to make and use these tools by watching their mothers. Different tool traditions are apparent in geographically isolated groups of chimpanzees: different types of tools are used for termite fishing at the Gombe in Tanzania and at sites in Senegal. Other culturally specific tool traditions exist; Goodall has never seen the Gombe chimpanzees use a stone as a hammer although there is an abundance of stones at Gombe. In contrast, Christophe and Hedwige Boesch (1981, 1984) have observed chimpanzees in a different part of Africa (Tai National Park, Ivory Coast) using stone tools to crack nuts open. During the nut season they spend on average two hours each day systematically gathering and cracking nuts, a rich source of food (an adult female gets about 4,000 calories, infants up to 1,000 calories a day from nuts). The nut-cracking technique is not mastered until adulthood, and at least four years of practice are necessary before any benefits are obtained. To open soft-shelled nuts they use thick sticks as hammers, with wood anvils. They crack harder-shelled nuts with stone hammers and wood anvils. These nuts have three segments, and the chimpanzee must rotate the nut on the anvil between each successive hammer blow to extract the whole kernel. Mothers overtly correct and instruct their infants from the time they first attempt to pound nuts, at age three years (Boesch and Boesch, in press).

The Tai chimpanzees live in a dense forest where suitable stones are hard to find. The stone anvils are stored in particular locations to which the chimpanzees continually return, and the wear patterns on the stones indicate that they have been used for generations. (Chimpanzees in their natural habitat appear to have a natural life span of about thirty years, which is not very different from that of people in preindustrial human societies. There is even a division of labor, with female chimpanzees specializing in nut cracking.) There is no conclusive evidence as yet that chimpanzees "make" stone tools by chipping away at stones to make them better hammers or anvils. However, such a finding would not be startling, given the fact that they "make" termite sticks and leaf sponges. Chimpanzees also use stones and other objects as projectiles with intent to do harm (Goodall, 1986).

Social Organization

Chimpanzees have what Goodall terms a "fusion-fission" society. They live in groups and know who members of the group are. This can be very important, because members of rival chimpanzee groups usually attack and sometimes even kill one another. Parts of the group split off to hunt, to mate, to "patrol" the boundaries of their range, or to engage in territorial warfare with other groups of chimpanzees. Several interesting features attend this social organization.

Warfare. Goodall documents a sequence of events that used to be regarded as unique to humans—warfare. The Gombe group split into two groups that for a time occupied two different areas. A stream seemed to be the recognized boundary for two years after the community split. However, after that interval each group mounted raids into the territory of the other group. If a stronger party with one or more males encountered a weaker "enemy" party it would attack with intent to kill, often using stones and other objects as projectiles. Over time the stronger group of chimpanzees killed their less numerous adversaries, who had once been their compatriots.

Sharing. Chimpanzee mothers share food with their children until they are about two or three years old, after which sharing decreases. Plant foods are never shared among adult chimpanzees. Meat, however, is shared; members of the group insistently beg meat from a chimpanzee who has been successful in a hunt.

Altruism, Compassion, and Moral Sense

Chimpanzees often come to the aid of other chimpanzees, usually close relations. However, Christopher Boehm (1981) has shown that such examples of altruism can be explained by the theory of kin selection (Hamilton, 1964). Acts that will preserve the lives of close relatives, who have many of the genes of the altruistic individual, may lead to the preservation of more of the individual's genes. For example, a person who dies while saving the lives of three of his sisters may transmit more of his genes to succeeding generations than if he had lived and they had died. Therefore, Darwinian natural selection can explain this sort of altruism, which I term *animal altruism* to differentiate it from human altruism. Chimpanzees also do not show the compassion for helpless individuals that marks some human societies. Although elder siblings will try to care for a young chimpanzee when its mother dies, infants are seldom adopted by unrelated females. Sick chimpanzees usually are shunned by other unrelated chimpanzees.

The difference between chimpanzee and human moral sense can be seen most clearly in the hunt. Chimpanzees, like many humans, hunt other animals to acquire meat. Goodall (1986) notes that a chimpanzee hunt is in many ways similar to that of humans: they cooperate and hurl projectiles at their prey, which usually are weaker animals (baboons, monkeys, piglets, antelope fawns). However, whereas human hunters kill their prey before they start to eat its flesh, chimpanzees do not seem to care whether the victim is dead or not. Typically they begin to eat small prey by biting open the skull, which causes death, but their purpose appears to be limited to keeping the victim immobile to facilitate the process of tearing into its flesh. The victim may scream and thrash about, but the chimpanzees' chief interest seems to be that the meal proceeds in an orderly manner, whether the victim is dead or alive. Goodall notes several examples:

> In 1977 Jomeo [a chimpanzee], after a brief struggle with an adult male colobus [monkey] in the trees, pulled his victim to the ground by the tail, then displayed, dragging the monkey. The colobus kept seizing hold of vegetation, but was not strong enough to break Jomeo's grip. His captor then let go and the monkey lay, squeaking faintly, but seemingly unable to walk. Adolescent Freud approached and bit at his hand, but retreated as the monkey moved violently. Five minutes later Jomeo's brother Sherry approached, took hold of the monkey's tail, dragged him a short way, let go, then turned around and bit at his face. Seizing the tail he displayed vigorously, flailing the monkey three times against trees and rocks. Jomeo immediately returned and also displayed, dragging and flailing the victim. Once again the monkey was left lying on the ground, still not quite dead. Another adolescent chimpanzee approached, stared, then bit off part of his genitals. Freud returned and bit off some more, and Jomeo completed the castration. At this point, nine minutes after his capture, the monkey was dead . . .

In 1980 Mustard chased and caught a female colobus with a fairly large infant. As he tried to seize her infant, he and the female fell to the ground and the infant escaped. The mother also ran off, hotly pursued by Mustard, who chased her for 20 meters or so, then yanked her tail and displayed with her . . . Mustard paused, seized her again and displayed vigorously, flailing her against the ground, then stamping on and hitting her. After this he sat down, turned the exhausted and battered female onto her back, and tried to bite into her belly. He flailed and screamed as she bit his hand. Once again he stood up and hit at her, then leaped onto her and stamped, still screaming loudly. His screams alerted Evered and Figan, who ran up and took over. Figan ripped open her belly, then bit at her face, while Evered tore off a leg; at this point she made her last sound and died.

Most of the baboon infants and juveniles whose deaths were observed died quickly. Often several adult male chimpanzees rushed up and tore the body of the victim apart. One ten-month-old infant, however, was consumed by only one adult male and was still alive and calling feebly for forty minutes after his capture. The three large bushpig young took between eleven and twenty-three minutes to die as they were slowly torn apart; the largest gave his final scream when Humphrey tore out his heart. (1986, pp. 291–292)

Language

Goodall and other close observers of chimpanzees believe that a combination of body gestures, facial expressions, and vocalizations serves as the chimpanzee's communications system. Goodall (1986, pp. 143–145) also notes that dialects exist in geographically isolated groups of chimpanzees, which argues for a communications system that is closer to human language than the fixed-call systems of simpler animals (Smith, 1977). The dialects that have been observed are nonverbal. For example, "At Gombe, when two chimpanzees are socially grooming, each often holds an overhead branch with one hand while grooming his or her companion with the other. At Mahale, 160 kilometers to the south, and in the Kibale forest of Uganda, a pair of grooming chimpanzees often sit, each clasping the head of the other at a point above their heads" (Goodall, 1986, p. 144).

New communications signals have been incorporated into the repertoires of the chimpanzee groups studied. For example, Shadow, a Gombe chimpanzee, invented and successfully used a novel courtship display (ibid., p. 145). Chimpanzees can identify friends and adversaries. They can call for aid or call to signal the presence of sources of food. They appear to be able to communicate both the nature and the location of sources of food, but the means employed are presently unknown (ibid., pp. 141–143). Chimpanzees, like other nonhuman primates, do not have voluntary control over their vocalizations. Their vocalizations seem to be bound into an emotional-display "package." For example, a chimpanzee who is about to feed will produce a cry even when doing so is counterproductive (ibid., p. 125). However, although the rough bounds of the chimpanzee communications system are known, how and what they communicate is presently a mystery.

We are in the curious situation of knowing more about what chimpanzees can do when they are exposed to human language than about their natural communications. Allen and Beatrix Gardner (1969, 1984) placed infant chimpanzees in a humanlike environment in which American Sign Language, ASL, was the means of communication. In contrast to the difficulties that chimpanzees have in producing the sounds of human speech, they can produce the manual gestures that specify ASL words. Their ASL words are not as clear as those of proficient adult humans who have acquired ASL as their first native language, but their signs are intelligible. Moreover, their ASL words have the properties of the words of human language in that they refer to concepts rather than to discrete items or events. The linguistic ability of ASL-trained chimpanzees seems to be about that of a two-and-a-half-year-old human, but they transmit words from one generation to the next without human intervention. The infant chimpanzee Loulis learned about fifty ASL signs from his five adolescent and adult chimpanzee companions (Fouts, Hirsch, and Fouts, 1982). The independent work of Sue Savage-Rumbaugh and her colleagues (1985, 1986) likewise shows that chimpanzees use and coin words. Chimpanzees also coin new words based on perceptual saliency or function. A brush used for tickling is identified by the ASL compound word "tickle-feather," analogous to the English word *toothbrush*; a duck is a "water-bird," analogous to a *steamboat*.

The ability of chimpanzees to acquire and use words should not be minimized. Words are inherently powerful aids to communication and cognition. As we communicate by means of language we are continually transmitting an implicit classificatory system, since words classify the phenomena that they code. When a child hears an adult use the word *tree* in reference to a tall leafy plant, a lesson is being taught. For example, the classificatory scheme that places dogs and cats into the superordinate category of domestic animals is transmitted to the child as she or he learns English in an American suburban household. In some other culture a chicken would fall into that category, but not a cat. The child goes on to expand the classificatory scheme when the names of various breeds of dogs are learned. Poodles, briards, and bulldogs are all dogs. In learning these words the child is acquiring the classificatory system shown in the diagram.

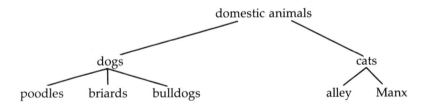

The inherent classificatory power of words probably explains, in part, the increased cognitive ability of bilingual children. Similar effects have been noted for language-trained chimpanzees who did better than other chimpanzees when they had to solve various sorts of problems (Premack and Woodruff, 1978).

Although these aspects of chimpanzee word use have been disputed (Terrace et al., 1979), it seems clear that chimpanzees have these abilities (Van Cantfort and Rimpau, 1982; Lieberman, 1984), and thus the potential ability to transmit culture. Several studies (Cheney and Seyfarth, 1980) of monkey calls and Goodall's (1986) observations of chimpanzee social organization suggest that nonhuman primates may use some sort of rudimentary words. However, present data on chimpanzees living in the wild do not show any evidence that clearly indicates that they use words.

What chimpanzees clearly do not seem to be able to acquire is the complex syntax of ASL. Careful observations of chimpanzee language using ASL and other modalities (such as sorting photographs or plastic symbols) show that syntactically significant word order is usually not preserved. A chimpanzee who by all appearances intends to sign *tickle me* is just as likely to sign *me tickle*. Recent data show that the pygmy chimpanzee (*Pan paniscus*) Kanzi at age five years does preserve word-order relations most of the time (Greenfield and Savage-Rumbaugh, in press). This result may indicate a species difference between "common" chimpanzees (*Pan troglodytes*) and *Pan paniscus*, or simply an individual difference between chimpanzees. It is difficult to determine whether this difference follows from Kanzi's being brighter than some other chimpanzees, since fewer than twenty chimpanzees have participated in all the language-training experiments to date.

Cognition

Chimpanzees have excellent mental maps; although they cannot be taught to read maps (Menzel, Premack, and Woodruff, 1978), they can remember over the course of a year the places in which food can be obtained at a particular season (Menzel, 1978). Chimpanzees also show intentionality. David Premack and Guy Woodruff (1978) showed videotaped "cliffhangers" to chimpanzees and had them complete the action portrayed. For example, they showed a person perched on top of a shaky set of boxes reaching for a banana. The chimpanzee could select a photograph that showed the person with the bananas in hand, or one that showed a pile of boxes and a person in disarray on the floor. When the video featured a laboratory assistant whom the chimpanzee liked, the banana feast was selected. A less favored laboratory assistant was matched with a fall.

Early Hominid Culture

Given our knowledge of what present-day apes can accomplish without all the attributes of human language and cognition, what inferences can we make about early hominid culture from the archaeological record? We can start by ruling out "lost" mental abilities.

A recurrent literary theme invests extinct hominids with abilities that are not found in present-day humans: telepathy, visualization of the inside of their bodies, and so forth. Obviously, we cannot take an australopithecine or Neanderthal into a laboratory and run a battery of cognitive tests, but we have tested chimpanzees. Chimpanzees do not have any cognitive abilities that are not better developed in humans; they do not perform better than humans do on any laboratory test (Premack, 1988). Since no special mental abilities are apparent over the range of hominoids from very advanced forms (people) to ones presumably close to the earliest hominids (chimpanzees), it is unlikely that any very different patterns of thinking existed in any extinct hominid.

Although this does not yield as romantic an image of the past as that of bands of hominids who were in some way better than us, it does allow us to evaluate the archaeological evidence of the culture of hominids who lived beyond the reach of our records and memories. If we assume that their minds probably worked in the same general way as ours, we can make some inferences about the late stages of hominid evolution. The evidence that we have consists of the stone and bone tools that they made, the remains of fires, scraps of bone and shell that show what they ate, a few traces of dwellings, some gravesites, beads, engraved bones, and some paintings and carvings, and their bones themselves. Cutting tools would have been used to carve the animals they hunted or scavenged, beads would have been used for adornment and possibly currency, paintings for rituals, and so forth.

Stone Tools and Toolmaking

The early hominids' stone tool technology seems to involve varying degrees of conceptualization. The earliest stone tools, dating from about 2.5 to 3 million years ago, were made by chipping away pieces from a stone that probably had the approximate shape of the intended finished object. The toolmaker had to keep in mind a conceptual image of the finished product. Chimpanzees do the same thing when they make termite sticks, though neither the technique used (stripping leaves from the branch) nor the conceptual target is very "abstract" in the sense of departing profoundly from the initial material—the branch with twigs and leaves.

More sophisticated was the Levalloisian "core-and-flake" technique, which became predominant 150,000 to 100,000 years ago (Bordes, 1968) (see Figure 6–1). An intermediate object, a turtle-shaped core, was first prepared by the chipping method. Then the core was struck with a hammer stone or pressed with a stick to shear a finished blade away at one stroke. The toolmaker had to keep in mind two conceptual stages, the intermediate core and the finished tool. Experiments with graduate students show that whereas one day and simple imitation and observation are enough to learn how to make the earlier type of simple tools, it takes more than one semester and specific

Figure 6–1. Two ways of making stone blade tools by the core-and-flake technique. After preparing the core the toolmaker can strike off a blade either by hitting a "punch" with a hammer or by gradually applying pressure.

instruction to master the core-and-flake technique (Washburn, 1969, p. 175). It is probable that having language enhances one's ability to teach someone else the core-and-flake technique, but it is difficult to say much more. Perhaps one way to determine whether human language is necessary to make complex stone tools by the core-and-flake technique would be to attempt to teach chimpanzees to make core-and-flake tools without explicit instruction using language.

A number of archaeological studies have attempted to develop metrics that reveal the cognitive complexity of various toolmaking techniques (for example, Bordes, 1968). However, Howard Dibble (1989) has pointed out that many of the traditional archaeological classifications of stone tools reflect the aesthetic complexities of the minds of the anthropologists who studied these tools rather than of the tools themselves. It seems to be impossible to say very much about the cognitive complexity of a toolmaker's mind on the basis of the tools that he or she made.

Furthermore, the relationship between the fossil record and tool technology does not show any close correlations. Earlier *Homo erectus* hominids had much simpler stone tools than did later Neanderthals and early modern humans. However, the stone tools used by both Neanderthals and anatomically modern humans living in the Middle East (for example, Shanidar, Qafzeh, and Skhul) do not differ in any significant way. They all are examples of the Mousterian tool culture, featuring core-and-flake techniques that once were ascribed to Neanderthals (Bordes, 1968). The modern human beings who replaced the Neanderthals in France 35,000 years ago used a more advanced stone tool technique, and until recently this advanced tool technology was often taken as an index of modern human cognitive capability. However, it is now evident that early modern *Homo sapiens* also used Mousterian tool technology throughout the period from 100,000 to 40,000 years ago. The Mousterian tool industry thus appears to be an earlier technology that was ultimately discarded by modern humans, but our ancestors used this technology for tens of thousands of years.

The lack of close detailed correlation between technology and cognitive ability should not be surprising. The tools that people use do not reflect their innate cognitive or linguistic ability in any direct manner. For example, we have no inherent cognitive advantages with respect to first-century Romans and Gauls even though the TGV high-speed train traveling between Paris and Nice at 200 kilometers per hour makes for a smoother ride than a chariot.[2] Various cultural factors can result in virtual technological stasis or sudden spurts. Relative complexity in tools and toolmaking technology is not a reliable indicator of the inherent cognitive capacity of the toolmakers.

Art

Artifacts that modern humans interpret as works of art do not occur until comparatively recent times, in conjunction with the fossil remains of anatomically modern humans. The earliest examples that are generally agreed to represent art are about 15,000 to 20,000 years old (Bordes, 1968). Alexander Marschack (1990) claims that some older European examples exist, dating from perhaps 45,000 years ago. The problem again is one of interpretation. What do these artifacts signify? Do they have religious significance? Are they decorative? In this connection body ornamentation appears to be a characteristic of anatomically modern humans. Body ornamentation appears in Western Europe about 35,000 years ago in association with modern human fossils (White, 1987). Decorative beads made of shells were often transported hundreds of kilometers and then painstakingly worked into their final form. Although the cognitive factors that are the bases of art may be very ancient, all present evidence is consistent with a fairly recent origin of art. Future archaeological research may lead to a reappraisal of the origins of art, but it is obvious that art cannot develop until the level of technology and culture reaches the point at which people do not have to spend all of their time in the struggle for existence.

Religious Thought and Moral Sense

One intriguing point raised by David Premack (1988) in connection with tests of the cognitive abilities of chimpanzees is how these abilities can be tested without the use of language. At some level it simply becomes impossible to run a test without using language. Consider the impossibility of determining whether dogs or chimpanzees can conceive of concepts such as life after death or reincarnation. It is impossible even to frame these questions without using human language at a level far above what can be grasped by a three-year-old.

Human religious thought and moral sense clearly rest on a cognitive-linguistic base. Therefore, archaeological evidence of ritual burials with grave goods indicates the presence of language and cognition beyond the base level represented by present-day chimpanzees. Burials in themselves may represent a mental state that transcends day-to-day life, reflecting "religious" beliefs. However, they may also reflect other concerns, such as surviving friends' and relatives' desire to see that the body is not torn apart by animals. Although this concern is one that nonhuman primates do not appear to share (or act on, since they cannot easily dig holes), it would not demonstrate the existence of any complex religious belief system. Burial likewise might reflect the desire of surviving members of the group to avoid the emotional trauma of continually seeing the remains of the corpse. Simple interment may represent the disposal of a cadaver in a depression along with other debris. The "disposal" of bodies by the Nazi SS death squads in Poland or by the Khmer Rouge in Cambodia falls into this category. The burials do not represent any religious beliefs on the part of the murderers or respect for the victims. Likewise, the burial of bodies as a public health measure in times of plague is not an expression of any belief in the hereafter and is not primarily an expression of respect for the dead. Therefore, we cannot assume that burial, in itself, represents a belief in the hereafter on the part of archaic hominids. In contrast, burials with grave goods clearly signify religious practices and concern for the dead that transcends daily life.

Although elaborate burials with clearly defined grave goods are comparatively recent, belonging to the Upper Paleolithic (about 35,000 years ago), the first evidence for burials with ritual grave goods occurs with the anatomically modern humans who lived 100,000 years ago. The people who lived in the Jebel

Qafzeh cave buried a child with ritual goods that were not part of the cave floor debris. The antlers of a fallow deer were placed in its hands (Vandermeersch, 1981). Several bodies were placed in either fully flexed or partially flexed positions in the extremely large (twenty-by-twelve-meter) Jebel Qafzeh cave. Skhul V, who like Qafzeh lived 100,000 years ago in what is today Israel, also was buried with ritual goods (McCowan and Keith, 1939); the mandible of a wild boar was placed in his hands, perhaps to signify prowess in the hunt in this life or to assure it in the next. A wild boar is a dangerous adversary even to hunters equipped with firearms. Skhul V and Qafzeh are two of the oldest fossil hominids who possessed the speech-producing anatomy and brains that are the biological bases of speech and syntax. The evidence of their burials with grave goods is consistent with their having possessed cognitive abilities that approach our own.

It is necessary to consider all the alternatives before interpreting particular archaeological data as evidence of religious rituals. Flower pollen was present in the Neanderthal grave found in Shanidar, Iraq, which may be about 60,000 years old; Ralph Solecki (1971) suggested that it had been placed in the gravesite as an offering. However, as Dibble (1989) notes, it is impossible to determine whether the flowers were deliberately placed in the grave or were simply part of the cave floor debris that fell into it. Similarly, the animal bones found in other burial sites may not have been deliberately placed near the graves. They may instead be part of the remains of meals littering the cave floor, which was also the living room and bedroom. However, a number of later Neanderthal burials (after 60,000 years ago) also appear to have included grave goods. The absence of earlier Neanderthal burials with grave goods may simply reflect the imperfection of the archaeological record, but a different interpretation is possible. The historical record shows that artifacts, techniques, and customs often are transferred from an advanced to a less advanced culture when they come into contact. Burial rituals incorporating grave goods may have been invented by the anatomically modern hominids who emigrated from Africa to the Middle East 100,000 years ago.[3] According to some theories, Neanderthal populations living in Europe and Asia moved southward to these regions during the glacial periods. Burial and other customs may have been transferred when these Neanderthal populations came into contact with the ancestors of modern human beings.

Although many present-day religious traditions do not include burial (Tibetan Buddhism, for example, mandates either cremation or "sky-burial," in which the body is chopped apart to be consumed by animals), some specific ritual is necessary to prepare the dead. However, burials are the only evidence that we have because they necessarily leave evidence behind. If we assume that the minds of our distant ancestors worked like ours, we can take burials that include grave goods as evidence for religious beliefs that predicate an afterlife, rebirth, or perhaps even reincarnation. Although there is a risk in assuming that the basic emotional needs of thinking humans have been similar over a span of 100,000 years, the alternative hypothesis—that profound differences exist—is even less likely. The continuity of evolution argues for the existence of similar emotional needs.

Selfless Behavior

Animal Altruism and Cooperation

It is obvious that altruism is not restricted to humans. The classic example is that of a worker bee's killing itself in the act of stinging an intruder to protect the hive. Goodall (1986) observed monkeys sacrificing their lives in attempts to save their infants. As Darwin (1859) noted, natural selection manifests itself in the behavior of animals as well as in their morphology. Our genes play a part in determining our behavior, and recent studies show that behavioral attributes such as shyness have a strong genetic component (Kagan, Reznick, and Snidman, 1988). And although cultural factors play an important part in fostering altruistic behavior, studies of altruism in humans show that it also has a strong genetic component. A study of 573 adult twin pairs showed that inheritance was the strongest factor in predicting altruistic and aggressive tendencies (Rushton et al., 1985).

Altruism might seem to be an unlikely candidate for evolution by Darwinian natural selection, since a person who gives food to another person or sacrifices himself seemingly will not do as well in the struggle for existence. However, natural selection operates by transmitting genes to the next generation and can account for the evolution of many aspects of altruistic behavior. As John Maynard-Smith explains:

> consider the fact that a parent may risk its life in defense of its offspring, say by feigning injury to distract a predator. In this way the parent may increase its own Darwinian fitness. Although it is possible that both the parent and offspring will be killed, it is more likely that both the parent and the offspring will survive. In the latter case the parent's Darwinian fitness will be greater after the altruistic act than it would have been if the parent had left its offspring to the predator. The genes associated with the altruistic act (in this case feigning injury) may be present in the offspring, so that their frequency may be increased. Hence natural selection favors parental altruism. (1978, p. 177)

William Hamilton (1964) first demonstrated that altruistic behavior to relatives as well as to offspring is a "winning" strategy. Studies using "game theory" show that various strategies that superficially do not appear to contribute to the survival of the animals performing them can lead to the transmission of more of their genes to future generations (Dawkins, 1976; Maynard-Smith, 1978; Parker, 1978). Cooperation or sharing, for example, often seems to be counterproductive to the cooperative individual. Why share food at all? The key to the evolution of cooperative behavior is a greater probability for the transmission of the genes of cooperating individuals to the next generation (Axelrod and Hamilton, 1981). Cooperation contributes to biological fitness through "tit-for-tat" interactions. An individual who helps another can expect help, or at least avoid aggressive negative acts, at some future time of need.

"Higher" Altruism

Jerome Kagan (1987) maintains that human altruistic behavior has an emotional basis. This is surely so insofar as the genetic expression of animal altruism undoubtedly involves the parts of the brain that regulate emotion. Like other social animals, we have brain mechanisms that cause us to act compassionately toward other genetically close animals—our relatives. Genetically programmed brain mechanisms may also foster cooperation in social groups to facilitate the tit-for-tat sharing of favors that promotes the survival of all members of the group. However, there are many examples of human altruistic behavior that simply cannot be explained by purely biosocial models, even ones that involve a rather extensive and general ledger sheet. St. Francis of Assisi's gift of his cloak to a leper is, for example, an altruistic act that does not contribute to biological fitness.

Altruism is at the core of human ethical and moral codes. The fourteenth Dalai Lama, for example, states that "the main theme of Buddhism is altruism based on compassion and love . . . In order to have a strong consideration for others' happiness and welfare, it is necessary to have a special altruistic attitude in which you take upon yourself the burden of helping others" (1984, pp. 32–33).

In my view, this "higher" human altruism evolved from human cognitive and linguistic ability acting on a preadaptive "emotional" base. This cognitive altruism is not restricted to interactions among individuals who are close relations. No increase in biological fitness necessarily accrues to an altruistic person. The preadaptive basis of cognitive altruism probably is the biologically based animal altruism discussed above, but we have made use of human cognitive ability to extend the concept of relation and we equally derive moral fitness rather than biological fitness by virtue of concepts that would be impossible to explicate without human language. In other words, human language and cognition are necessary conditions for moral sense.

The teachings of the fourteenth Dalai Lama illustrate this claim. The Tibetan Buddhist argument for the basis and attainment of altruistic behavior is based on (1) concepts that inherently would be impossible to state without human language and (2) reason. The premises of Tibetan Buddhism derive from the fact that life on earth is generally painful and tumultuous. Buddhist doctrine predicates an endless series of rebirths from which one can escape only by achieving a correct attitude and practicing good works. Altruism predicated on compassion and love is one of the key elements in achieving the state of nirvana—a state of perfection and release from the cycle of rebirth; the other is a sense of the self, discussed in the next pages. However, altruism itself is not an innate automatic quality; it can be achieved only by training and thought:

> . . . you must first train in a sense of kindness through using as a model a person in this lifetime who was very kind to yourself and then extending this sense of gratitude to all beings. Since, in general, in this life your mother was the closest and offered the most help, the process of meditation begins with recognizing all other sentient beings as like your mother.
>
> Since rebirths are perforce infinite, everyone has had a relationship with yourself like that of your own mother of this lifetime . . . Since our births are beginningless, there is no limit to their number; thus it is not definite that those who are now our friends were always friends in the past and that those who are now our enemies were always enemies in the past . . . Hence there is no sense in one-pointedly considering a certain person to be just a friend and another person to be just an enemy . . . the next step is to consider that since everyone's births have been beginningless and thus limitless in number, every single person has been your best of friends, parent or whatever, over the course of lifetimes. Taking this realization as a basis, you can slowly develop an attitude considering all sentient beings to be friends.
>
> Then, consider the kindness that they individually afforded to you when they were your parents. When they were your mother or father, usually the best of all friends, they protected you with kindness just as your parents did in this lifetime when you were small. Since there is no difference in the fact people have been kind to you whether they expressed that kindness recently or a while ago, all beings have equally shown kindness to you either in this lifetime or in other lifetimes; they are all equally kind. (Dalai Lama, 1984, p. 35)

The key elements in the Dalai Lama's argument thus constitute a logical argument: (1) Endless rebirth is a fact. (2) Therefore, anyone you meet *was* your mother or father in some previous life because everyone has been reborn an infinite number of times. (3) This being the case, you should treat everyone as you would treat your closest biological relatives. Although altruism may derive from a genetic basis resulting from natural selection, the higher altruism that is at the ethical core of Buddhism has a cognitive-linguistic basis. It must be carefully studied and nurtured; the point of Tibetan Buddhist theology and instruction is the attainment of the correct altruistic attitude.

Other aspects of Tibetan Buddhist doctrine also are predicated on the prior existence of human cognitive and linguistic ability. Buddhism assumes the existence of different levels of cognition. The concept of karma involves a kind of a ledger sheet in which acts and attitudes that bring merit to or detract from it are noted; the ledger sheet is balanced at the end of one life and determines the nature of one's next rebirth (Dalai Lama, 1984). Buddhism is not unique among the systems of ethical and religious thought that attempt to guide human behavior. Kant, for example, in his *Groundings for the Metaphysics of Morals*, attempts "to draw these concepts and laws from pure reason, to present them pure and unmixed, and indeed to determine the extent of this entire practical and pure rational cognition, i.e., to determine the whole faculty of pure practical reason" (1981 [1785], p. 23).

Although altruistic behavior in human children seems to derive from an innate, genetic component, it is fostered by early experience. Virtually all the children who have been studied by Carolyn Zahn-Wexler, Barbara Hollenbeck, and Marian Radke-Yarrow (1984) showed empathy toward other people and pets before age two. However, their behavior was not very different in this respect from that of the pet dogs who lived in their homes. The earliest stages of humanlike altruistic behavior occur at about the same age as language and cognition begin to be evident. Zahn-Wexler and her colleagues note that

> just a little past the first year of life, children begin to comfort others in distress. This is a developmental landmark; an aversive experience in another person draws out a concerned, approach response from the child. Children's first prosocial acts are physical interventions: they pat and hug the victims, rub their hurts

and so on . . . There is also an explosion of prosocial activity at this time. Children's acts of compassion begin to take many different forms: acts of help, sharing, comforting, rescue, distraction, defense/protection, verbal sympathy, are now present . . . Virtually all of the children studied showed this early concern for the welfare of another being. This *uniformity* suggests that altruism is a biological given, "wired" in and ready for expression given sufficient physical, cognitive and emotional growth. (Zahn-Wexler et al., 1984, p. 29)

Generalized altruism is, in the view of these researchers, a learned behavior that builds on this genetically transmitted base. Cognitive development is not enough; altruism must be taught and is in no sense an automatic consequence of simple exposure to abstract principles or isolated examples of altruistic behavior. Zahn-Wexler and her colleagues conclude that

the parent who is altruistic to others but is cold with his child is not going to have much success in developing generalized altruism in his child. Further, the parent who conveys his moral values as principles only, but does not translate these into real, caring actions, accomplishes a similar limited kind of learning in the child. Generalized altruism appears to be best learned from parents who both inculcate the principles and show real altruism in their everyday interactions. And their practices toward their children are consistent with their general altruism. (Zahn-Wexler, Hollenbeck, and Radke-Yarrow, 1984, p. 34)

Children raised in abnormal circumstances do not develop normal general altruistic or social behavior. For example, the children of manic-depressive parents, who are unable to provide a consistent emotional environment, have difficulty in engaging in friendly play with other children, in sharing with their friends, and other aspects of altruistic behavior (Zahn-Wexler et al., 1984). Clearly, altruism in the higher human sense is a culturally transmitted behavior that has both a cognitive and an affective base.

Animal versus Cognitive Altruism

There is a conflict between the sort of animal altruism that is biologically transmitted and higher human altruism. The functional value of animal altruism derives from the transmission of one's genes to future generations. Thus altruistic behavior to unrelated persons can be counterproductive in the absence of a tit-for-tat explicit or implicit bargain. For example, feeding songbirds or starving wild deer (who cannot be hunted) in winter is a counterproductive act at the animal-altruism level. Moreover, certain acts that are morally repugnant are "useful" at the animal-altruism level. The wartime activities of the Nazi SS fall into this category. Hundreds of innocent civilian hostages were routinely murdered when German troops were attacked by partisans; the mass murders terrorized the population, discouraging future attacks. These murders were directed toward preserving the genes of the SS's presumed "closer" genetic relatives—the members of the "master race," thereby increasing their putative biological fitness. If we were simply enhancing our biological fitness, we would simply push into the sea all the persons in a lifeboat who were not close relations.

True altruism involves a cognitive act, considering the "other" as your mother if you are a practicing Buddhist, or putting yourself in the place of the other. The Golden Rule of Maimonides, "Do unto others as you would have done to yourself," necessarily involves the ability to recognize the self and then transpose your self to some other being's viewpoint. The fourteenth Dalai Lama, for example, explains that to surmount selfish behavior

> one should visualize the following: On one side imagine your own "I" which so far has just concentrated on selfish aims. On the other side imagine others—limitless, infinite beings. You yourself are a third person, in the middle, looking at those on either side. As far as the feeling of wanting happiness and not wanting suffering are concerned, the two sides are equal, absolutely the same. Also with regard to the right to obtain happiness they are exactly the same. However, no matter how important the selfishly motivated person is, he or she is only one single person; no matter how poor the others are, they are limitless, infinite. The unbiased third person naturally can see that the many are more important than the one. Through this, we can experience, can feel, that the majority—the other limitless be-

ings—are more important than the single person "I." (1984, p. 11)

Human children do not achieve the highest levels of pretend play, in which they put themselves into the role of some other person or being, until about age five. Moral behavior equivalent to that of an adult begins to be evident at this age (Fischer, 1980; Kagan, 1987). Cognitive altruism has probably evolved in human culture because, as with other aspects of evolution, there is a selective advantage for the individual who adopts this pattern of behavior, but the advantage is in the abstract realm of thought. Cognitive altruism increases an individual's cognitive fitness. For a person who holds a belief with a cognitive basis, such as the attainment of Nirvana or heaven, altruism increases the likelihood of attaining that goal. Without this cognitive basis, neither belief in or enhancement of one's chances of attaining heaven or Nirvana is possible.

These moral and ethical systems appear to be both recent and fragile. The great world religions are all less than 10,000 years old, and their precepts are frequently violated on a massive scale. We know that the shamanistic systems that preceded Christianity in Europe and Buddhism in Asia were barbarous and cruel. Human sacrifice appears to have been a feature of both druidism in England and Bon-Po shamanism in Central Asia. However, to the extent that a belief in other worlds demonstrates the linguistic and cognitive prerequisites for higher moral sense, evidence for any system of religious belief is germane to the question of the evolution of moral sense. Archaeological evidence of such beliefs, in the form of ritual burial and grave goods, is thus crucial. These practices appear to have evolved in the last 100,000 years, although they were not really well developed until the beginning of the Neolithic, which occurred in the Middle East 12,000 years ago.

On Being Human

The earliest fossil remains of anatomically modern *Homo sapiens* that can be accurately dated lived about 100,000 years ago in what is today Israel. These early humans had modern supralaryngeal vocal tracts and the brain mechanisms that are necessary to produce human speech and syntax. They probably had a language, or languages, that made use of complex syntax and reasoning ability. They thought and talked about death and constructed theories about life, death, and a world that might exist beyond death—for which they prepared their dead in burial. Human language and thought may be still older. If the dating established by recent studies of DNA is accurate, modern human beings may have first appeared 250,000 years ago; and it is possible that they also talked, thought, and acted in this manner, though perhaps to a reduced degree. However, we can date language as we know it back to at least this period and place—100,000 years ago at the edge of Africa and Asia. And we can trace the human drive to construct theories and codes concerning our place in life, the meaning of death, and the conduct of life—that is, religious and moral systems—to these people.

The development of human culture, of which moral sense is arguably the highest form, has obviously progressed in the last 100,000 years. We can see progress in the last century. Slavery has been almost universally outlawed; torture is concealed by most governments because it is now considered unacceptable and abhorrent. However, it is also obvious that our moral development is imperfect. Although we have populated and changed the continents, harnessed the forces of nature, and subjugated every other form of life, we have not conquered ourselves. The remnants of the primitive brains within ourselves still generate the rage, anger, and violence that dominate human affairs. If these behavioral attributes—altruism, empathy, and moral sense—are markers of fully modern human beings, then it is apparent that this aspect of our potential is still incomplete. Our only hope is to use our unique evolutionary heritage—the powers of human language and human thought—in the service of selfless behavior and moral sense.

Notes / *References* / *Index*

Notes

1. Brain Structure, Behavior, and Circuitry

1. Aquatic mammals such as dolphins also have very large brains, but these brains have a very different, largely undifferentiated cortical structure. Although dolphins have a complex communications system and are clever, their behavior does not even start to approach that of human beings (Schusterman and Gisiner, 1988). Their large brain probably is involved with their ability to use sounds underwater to localize objects—a biological SONAR system.

2. Some experiments have reported that the supplementary motor area of the monkey neocortex has some effects on vocalization (Sutton and Jurgens, 1988), but the general control of vocalization is quite different from that in human beings. See Chapter 4.

3. The words *higher* and *lower* should not be taken as endorsing the directionality of evolution. With regard to the evolution of the brain, animals that are closely related and evolved from some common ancestor are *higher* if they evolved later and generally have more complex brains.

4. Some cytoarchitectonic studies suggest that area 43 and areas 44 and 45 (Broca's region) are more similar to the structure of the prefrontal cortex, but these studies are disputed. Moreover, areas 44 and 45 are functionally premotor (Stuss and Benson, 1986, pp. 15–16).

5. As time goes on, computer simulations of distributed neural networks are incorporating various aspects of real brains. For example, some recent models incorporate synaptic inhibition, so that the transmission of a signal from one neuron to another can be impeded (Bear, Cooper, and Ebner, 1987). Edelman's (1987) model operates with *groups* of neurons that correspond to the columns of neurons that can be observed in the cortex. Edelman's model furthermore forms a pattern of permanent interconnections between neurons that reflects the "experience" of the distributed neural network; "genetically" identical

computer-simulated neural networks therefore become quite different as the result of different "life experiences." This effect again corresponds to the behavior of biological brains.

2. Human Speech

1. The breath-groups have the basic form of the mammalian isolation cry discussed in Chapter 1. They usually segment the flow of speech into sentence-length segments, but they can be optionally used to segment a message into the "phrases" that constitute the sentence. An adult speaker, for example, will usually produce the sentence "Bill and Jerry saw the silly old man" in one breath-group, but the sentence could be produced in two linked breath-groups "[Bill and Jerry] [saw the silly old man.]" Young children can use these intonational cues to master the syntactic structure of their native language (see Chapter 4 and Gleitman et al., 1987). The linguistic uses of the breath-group constitute an interesting example of Darwinian preadaption, using an old mechanism for new ends.

2. Johannes Müller (1848), who first formulated the source-filter theory of speech production, also was the first scientist to pose the question of *why* these sounds occurred more often in different languages. Müller stated that it was the province of physiology to answer this question.

3. The general enlargement of the brain is, as was noted in Chapter 1, a significant factor in hominid evolution that can be traced in the fossil record. However, it is almost impossible to determine the details of brain organization from the fossil record. Holloway (1985), who has studied the "endocasts" that can be made of the outer surface of the brain, using the inside of the skull as a reference, concludes that the uncertainties preclude definite conclusions at this time. For example, although Neanderthal hominids had brains that are as large as those of modern human beings, endocasts do not show whether they were modern or not. See the next chapter.

4. The dating of the Broken Hill fossil is presently uncertain, and it is possible that anatomically modern *Homo sapiens* first appeared in the Middle East 100,000 years ago. However, molecular evidence points to an African origin; accurate dates for early anatomically modern fossils in Africa would resolve this debate.

3. A Thoroughly Modern Human Brain

1. We can make some reasonable inferences regarding australopithecine culture by using the known behavior of present-day chimpanzees as a baseline for early hominid behavior. Although chimpanzees are not

early hominids, they are much closer to the common ancestor of early hominids and apes than are modern human beings. The cultural attainments of chimpanzees, which involve toolmaking and complex social interaction, probably were shared by australopithecines. Chapter 6 discusses these questions in greater detail.

2. At least one language, Warlpiri, which is spoken in Australia, allows a speaker to vary the order in which words appear within a sentence (Bavin and Shopen, 1985). If Bavin's analysis is correct, the complex "universal grammar" that Chomsky (1975, 1976, 1980a, 1980b, 1986) claims typifies *all* human languages would appear to be irrelevant for Warlpiri. Many of Chomsky's adherents dispute Bavin's findings and claim that Warlpiri has syntactic rules that do not differ from languages such as English, in which the sequence in which the words of a sentence may appear is constrained. However, this dispute is irrelevant to the question that we must address; even if some languages happen to lack syntax (through some accidents of history), we still must account for the syntactic ability of human beings. Bavin, for example, points out that Warlpiri children, given appropriate social conditions, have no difficulty mastering English.

3. Although some studies have stated that lesions in the supplementary motor area of the monkey neocortex can affect their vocalizations, recent data show that this is not the case (MacLean and Newman, 1988).

4. Different terms and systems for classifying the effects of brain damage on speech and language have been proposed. Some of these systems implicitly make different claims concerning the nature of aphasia (see Caplan, 1987, for a comprehensive discussion). However, the term *Broca's aphasia* has historical priority and will be retained even though it is now evident that these deficits do not follow from damage localized to Broca's area (Brodmann areas 44 and 45).

5. The term *aphemia* is used here in the restricted sense proposed by Stuss and Benson (1986); the patient is initially mute or may have profound speech production problems but then recovers. In contrast, Broca (1861) used the term *aphemia* to describe deficits in motor control that persisted and appeared to be independent of any other language deficits. The term *apraxia* is now usually used to describe this condition (Darley, Aronson, and Brown, 1975).

6. Baum tested eight agrammatic aphasic and six age-matched normal subjects using both a task that tests the ability of a subject to use grammatical information instantly (i.e., automatically) while listening to a sentence, and a grammatical decision test in which the subject can use more deliberate, introspective "controlled" strategies (Posner and Snyder, 1975). The aphasics' performance differed from that of normal control subjects on both tasks: they were unable to access the

grammatical rules of English automatically; and they *also* performed poorly when making grammaticality judgments, with an error rate of 29 percent compared with almost no errors for the normal controls.

7. Poizner, Klima, and Bellugi (1987) claim that the signing deficits of the aphasics whom they studied were limited to ASL, i.e., that they did not occur when the subjects were asked to make nonlinguistic hand gestures. They conclude that ASL involves a "module" independent from manual control. However, Kimura (1988) disputes this claim. She notes other studies showing that deaf aphasics consistently have difficulty in executing nonlanguage hand gestures equal in complexity to those used in ASL. Kimura also disputes the validity of the basic data used by Poizner, Klima, and Bellugi, since she happened to have previously tested one of the subjects used in their study and found that, contrary to their claims, he had difficulty in executing nonlinguistic hand gestures (Kimura, Battison, and Lubert, 1976).

8. They also note that Wernicke's aphasia may derive from lack of information from the temporoparietal cortex to normally functioning prefrontal regions. Their data also demonstrate that conduction aphasia cannot be explained by the Geschwind (1964) model, since the PET scans show that "inputs to the frontal regions from posterior language regions are not diminished" (p. 33).

9. Some neurologists have claimed that the dementia associated with subcortical disease derives from Alzheimer's disease. However, the data discussed later in this chapter show that the dementia associated with subcortical disease is quite different from that associated with Alzheimer's disease, in which speech and syntax are preserved (Kempler, Curtiss, and Jackson, 1987).

10. The CT scans of the two agrammatical deaf ASL aphasics studied by Bellugi, Poizner, and Klima (1983) also are consistent with this hypothesis. The prefrontal cortex was either directly damaged (subject G. D.) or disconnected from Broca's area as a result of subcortical damage (subject P. D.).

11. The role of the basal ganglia in language and thought in the circuit theory presented here differs from that hypothesized by Jason Brown (1988). The claim here is that these structures have become modified by the process of evolution and play a direct role in making human language and thought possible. In contrast, Brown has proposed that brain mechanisms such as the basal ganglia operate in the human brain in much the same way that they operate in simpler animals. According to Brown, higher human behaviors such as language and thought involve the neocortex's suppression of the activity of more primitive mechanisms such as the basal ganglia.

12. This discussion follows the definitions and terminology of André Parent (1986).

13. Alexander, Naeser, and Palumbo (1987) claim that extensive damage to another subcortical structure, the periventricular white matter (PVWM), may cause the speech deficits associated with subcortical lesions. However, their own data show that this is not always the case; patient 5 in their study had extensive damage to the putamen and internal capsule but only a small lesion in the PVWM. His "speech was very abnormal, with articulation, prosody and volume all impaired . . . comprehension was impaired at the level of complex material or syntax-dependent material" (p. 966). As Alexander notes, there are probably redundant alternate subcortical pathways involved in speech production that complicate the picture.

14. The "motor theory of speech perception," which claims that humans always interpret speech signals in terms of underlying articulatory movements (Liberman et al., 1967; Liberman and Mattingly, 1985), would appear to furnish a basis for the lateralization of speech perception. However, it is often the case that rather different acoustic signals correspond to the same speech sound—e.g., the different formant frequency patterns generated for the "same" vowel by speakers who have different length vocal tracts. According to the classic motor theory of speech perception, only one articulatory maneuver could generate these varied acoustic signals; therefore, a process of internal reconstruction would account for a listener's interpreting all these different acoustic signals as tokens of the same articulatory maneuver executed by different vocal tracts. If this were true, the lateralized aspects of the human brain that regulate the production of speech might also be involved in its perception. However, a strict interpretation of the motor theory of speech perception is not tenable. Although some speech sounds such as the vowel [i] are always produced with the same articulatory maneuvers, other sounds can be produced by means of different articulatory maneuvers (Ladefoged et al., 1972). The acoustic signals that specify the sounds of speech therefore cannot be consistently decoded into invariant articulatory maneuvers. Indeed, as we saw in Chapter 2, humans' ability to use alternate articulatory maneuvers to achieve the same acoustic signal is one of the interesting characteristics of the automatized control of human speech. The brain mechanisms that control human speech apparently "know" an entire set of different articulatory maneuvers that in different circumstances yield the same acoustic signal.

15. Deacon (1990) claims that the brain circuits of classic Neanderthal hominids were identical with those that occur in the human brain today. Deacon bases his argument on his tracer studies (1988a) of the brains of monkeys. The geometry of certain pathways in the general region of the monkey homologue to Broca's area is similar to that indicated by electrical stimulation studies in humans (Ojemann, 1983).

Deacon argues that the similarity of these pathways shows that the language circuitry of human brains is similar to that of monkeys; hence Neanderthal brains would have the same circuitry and, being as large as human brains, would be as well adapted for speech and language. However, behavioral and neurophysiologic data show that monkey brains *cannot* control speech production. Therefore, we can conclude that the circuits that Deacon has identified are *not* the ones that differentiate the human from the monkey brain. Subcortical circuits are probably involved in the human-monkey distinction.

4. The Brain's Dictionary

1. The acoustic signal that the parrot produces when it "talks" is not equivalent to that of human speech because birds inherently cannot produce the sounds of human speech (Greenewalt, 1968). Human listeners interpret the parrot's acoustic signal as speech although it is distorted.
2. As noted by Patricia Churchland.

5. Learning to Talk and Think

1. According to what might be termed the "MIT school of linguists," the Universal Grammar is responsible for only the "core" of a language's syntactic rules. The rest of the language constitutes the "periphery," which a child presumably learns by means of general cognitive mechanisms. The core/periphery distinction is, as James McCawley (1988) notes, a very "shaky" one that makes the theory virtually untestable.
2. The Universal Grammar determines only the characteristics of the "core language." The core language is not the language that a person actually uses. It is "a system determined by the parameters of UG [Universal Grammar]"; the "periphery," the language actually used, "is added on in the system actually represented in the mind/brain of a speaker-hearer" (Chomsky, 1986, p. 147). The periphery is presumably learned by means of general cognitive mechanisms, because it is not specified in the Universal Grammar. James D. McCawley points out a problem with the core/periphery distinction and the Universal Grammar. If we accept this distinction, then the language learner must use general cognitive mechanisms to learn the many aspects of language that are in the putative periphery. If this is the case, why wouldn't they continue to use these general learning methods for the "core"? As McCawley puts it, "it is a mistake to assume that because human beings have some biologically determined mechanisms specific to language acquisition, those mechanisms do the whole job . . . Since by Chomsky's definition, the periphery of things that can't be

acquired by just the setting of parameters [the UG], it must be acquired by some other mechanisms. Until advocates of the core/periphery distinction come to some conclusions as to how the periphery is acquired, they are in no position to claim that the same machinery by which the periphery is acquired does not also play a role, perhaps even a major one, in the acquisition of the core. Here as before, the core/periphery distinction is on shaky grounds" (1988, p. 154).

3. For example, Elan Dresher (1989) has proposed a set of eleven principles and five sequential constraints that supposedly are necessary for children to be able to pronounce correctly the stress patterns of the words of their native language. Dresher has constructed an ingenious interlocked set of nonredundant rules. If any of these sixteen items were not available to the child, she or he supposedly would never be able to learn to pronounce words. A more probable set of "innate stress acquisition rules" would consist of loosely specified, redundant genetically transmitted information concerning possible stress patterns. The acquisition of a child's first language would then involve the interplay of general cognitive ability and the innate language-specific information loosely coded in the brain's Universal Grammar.

4. Normal interaction apparently requires social interaction using language. Catherine Snow (1977) studied Dutch children who heard lots of German while their parents viewed German TV shows. The parents did not speak to their children in German, and the children did not "acquire" German. This finding contrasts markedly with the "activation" of a genetic blueprint for bird song that has been documented by Marler (1976) and Nottebohm (1984). Songbirds who are briefly exposed to species-specific songs acquire these songs even when isolated from other conspecific birds. The birds act as though they had a "universal song grammar."

5. Cuing effects occur when a subject responds to the experimenter's face, body, and posture rather than to the question that is being posed. The horse "Clever Hans" supposedly was able to add numbers together by moving his foot. For example, he would move his hoof seven times when he was asked to add 4 and 3. What was happening was that Hans had noticed that his trainer's facial expression changed when Hans came to the correct solution. Since Hans was rewarded for each successful arithmetic operation, he stopped pawing when this unconscious facial movement occurred (Pfungst, 1907). Many experiments that ostensibly test the cognitive abilities of children are contaminated by cuing effects. For example, in some recent experiments children were asked to tell the experimenter, who knows English, whether a sentence was grammatical, that is, "good," while the experimenter was in full view of the child. In these conditions it is

almost certain that the experimenter's face and demeanor will provide as much information to the child as the structure of English.
6. Gleitman et al. (1987) point out various "general" cognitive processes that are necessary to determine the particular form of a language *even* if a "universal grammar" exists that determines the general properties of all human languages.

6. Culture and Selfless Behavior

1. Present-day chimpanzees are, of course, neither ancient hominids nor the ancient apes who were their ancestors; they have evolved over the past 3 to 6 million years.
2. The complexity of a culture is not necessarily evidenced by its tools. For example, the military tactics and organization of the Roman legions were far more complex and effective than those of their barbarian opponents, but their weapons did not materially differ.
3. Some biological anthropologists claim that modern human beings evolved in the Middle East rather than in Africa. This hypothesis is less likely than an African origin, but it would not bear on this argument.

References

Albert, M. A., R. G. Feldman, and A. L. Willis. 1974. The "subcortical dementia" of progressive supranuclear palsy. *Journal of Neurology, Neurosurgery, and Psychiatry* 37:121–130.

Alexander, M. P., M. A. Naeser, and C. L. Palumbo. 1987. Correlations of subcortical CT lesion sites and aphasia profiles. *Brain* 110:961–991.

Altman, J. 1987. Cerebral cortex: A quiet revolution in thinking. *Nature* 328:572–573.

Anderson, A. 1988. Learning from a computer cat. *Nature* 331:657–659.

Anderson, J. A. 1988. Concept formation in neural networks: Implications for evolution of cognitive functions. *Human Evolution* 3:83–100.

Arensburg, B., A. M. Tiller, B. Vandermeersch, H. Duday, L. A. Schepartz, and Y. Rak. 1989. A middle Paleolithic human hyoid bone. *Nature* 338:758–760.

Armstrong, L. E., and I. C. Ward. 1926. *Handbook of English intonation.* Leipzig and Berlin: Teubner.

Atkinson, J. R. 1973. Aspects of intonation in speech: Implications from an experimental study of fundamental frequency. Ph.D. diss., University of Connecticut.

Au, T. K. 1983. Chinese and English counterfactuals: The Sapir-Whorf hypothesis revisited. *Cognition* 15:155–187.

Axelrod, R., and W. D. Hamilton. 1981. The evolution of cooperation. *Science* 211:1390–96.

Bachrach, J. A., and R. L. Karen. 1969. *Complex behavior chaining* (film). University Park, Pa.: Psychological Film Register.

Bain, B., and A. Yu. 1980. Cognitive consequences of raising children bilingually: One parent, one language. *Canadian Journal of Psychology* 34:304–313.

Baldwin, J. D., and J. I. Baldwin. 1977. The role of learning phenomena in the ontogeny of exploration and play. In *Primate bio-social development: Biological, social and ecological determinants*, ed. S. Chevalier-Skolnikoff and F. E. Poirer, 343–406. New York: Garland.

Baum, S. R. 1988. Syntactic processing in agrammatism: Evidence from lexical decision and grammaticality judgment tasks. *Aphasiology* 2:117–135.

Baum, S. R., S. E. Blumstein, M. A. Naeser, and C. L. Palumbo. In press. Temporal dimensions of consonant and vowel production: An acoustic and CT scan analysis of aphasic speech. *Brain and Language.*

Bavin, E. C., and T. Shopen. 1985. Children's acquisition of Warlpiri. *Journal of Child Language* 12:597–601.

Bayles, K. 1984. Language deficits in Huntington's and Parkinson's disease. Lecture delivered to the Academy of Aphasia, Los Angeles.

Bayles, K., and D. R. Boone. 1982. The potential of language tasks for identifying senile dementia. *Journal of Speech and Hearing Disorders* 47:210–217.

Bayles, K., and C. K. Tomoeda. 1983. Confirmation naming in dementia. *Brain and Language* 19:98–114.

Bear, M. F., L. N. Cooper, and F. F. Ebner. 1987. A physiological basis for a theory of synaptic modification. *Science* 237:42–48.

Bellugi, U., H. Poizner, and E. S. Klima. 1983. Brain organization for language: Clues from sign aphasia. *Human Neurobiology* 2:155–170.

Benson, D. F., and N. Geschwind. 1972. Psychiatric conditions associated with focal lesions of the central nervous system. In *American handbook of psychiatry,* ed. M. F. Reiser. New York: Basic Books.

Bloom, L. 1973. *One word at a time: The use of single-word utterances before syntax.* The Hague: Mouton.

——— 1985. Aphasia and related disorders: A clinical approach. In *Principles of behavioral neurology,* ed. M. M. Mesulam, 193–228. Philadelphia: F. A. Davis.

Ben-Zeev, S. 1977. The effect of bilingualism on cognitive strategy and cognitive development. *Child Development* 48:1009–18.

Berlin, B., and P. Kay. 1969. *Basic color terms: Their universality and evolution.* Berkeley: University of California Press.

Bloom, A. H. 1981. *The linguistic shaping of thought: A study of the impact of language on thinking in China and the West.* Hillsdale, N.J.: Lawrence Erlbaum Associates.

Bloom, L. 1973. *One word at a time: The use of single-word utterances before syntax.* The Hague: Mouton.

Blumstein, E. E. 1981. Neurolinguistics: Language-brain relationships. In *Handbook of clinical neurophysiology,* ed. S. B. Filskov and T. J. Boll, 227–256. New York: Wiley.

Blumstein, S. E., W. Cooper, H. Goodglass, H. Statlender, and J. Gottleib. 1980. Production deficits in aphasia: A voice-onset time analysis. *Brain and Language* 9:153–170.

Blumstein, S. E., and K. N. Stevens. 1979. Acoustic invariance in speech production: Evidence from measurements of the spectral properties of stop consonants. *Journal of the Acoustical Society of America* 66:1001–17.

Boehm, C. 1981. Parasitic selection and group selection: A study of conflict interference in rhesus and Japanese macaque monkeys. In *Primate behavior and sociobiology*, ed. A. B. Chiarelli and R. S. Corruccini, 161–182. Berlin: Springer-Verlag.

Boesch, C., and H. Boesch. 1981. Sex differences in the use of natural hammers by wild chimpanzees: A preliminary report. *Journal of Human Evolution* 10:585–593.

────── 1984. Possible causes of sex differences in the use of natural hammers by wild chimpanzees. *Journal of Human Evolution* 13:415–440.

────── In press. Tool use and tool making in wild chimpanzees. *Folia Primatologica*.

Bond, Z. S. 1976. Identification of vowels excerpted from neutral nasal contexts. *Journal of the Acoustical Society of America* 59:1229–32.

Bordes, F. 1968. *The old stone age.* New York: McGraw-Hill.

Bouhuys, A. 1974. *Breathing.* New York: Grune and Stratton.

Boule, M. 1911–1913. L'homme fossile de La Chapelle-aux-Saints. *Annales Paléontologie* 6:109; 7:21, 85; 8:1.

Boule, M., and H. V. Vallois. 1957. *Fossil men.* New York: Dryden Press.

Bowerman, M. 1987. What shapes children's grammars? In *The cross-linguistic study of language acquisition*, ed. D. I. Slobin. Hillsdale, N.J.: Lawrence Erlbaum Associates.

────── 1988. The role of meaning in grammatical development: A continuing challenge for theories of language acquisition. In *Proceedings of the 13th Annual Boston University Conference on Language Development.* Vol. 7. Boston: Program in Applied Linguistics, Boston University.

Bradshaw, J. L., and N. C. Nettleton. 1981. The nature of hemispheric lateralization in man. *Behavioral and Brain Sciences* 4:51–92.

Broca, P. 1861. Remarques sur le siège de la faculté de la parole articulée, suivies d'une observation d'aphémie (perte de parole). *Bulletin de la Société d'Anatomie* 36:330–357.

Brodmann, K. 1908. Beiträge zur histologischen Lokalisation der Grosshirnrinde. VII. Mitteilung: Die cytoarchitektonische Cortexgleiderung der Halbaffen (Lemuriden). *Journal für Psychologie und Neurologie* 10:287–334.

────── 1909. *Vergleichende histologische Lokalisation der Grosshirnrinde in iheren Prinzipen Dargestellt auf Grund des Zellenbaues.* Leipzig: Barth.

────── 1912. Ergebnisse über die vergleichende histologische Lokalisation der Grosshirnrinde mit besonderer Berucksichtigung des Stirnhirns. *Anatomischer Anzeiger* (Suppl.) 41:157–216.

Bronowski, J. 1978. *The origins of knowledge and imagination.* New Haven: Yale University Press.

Brooks, V. B. 1986. *The neural basis of motor control.* New York: Oxford University Press.

Brown, J. W. 1988. *The life of the mind: Selected papers.* Hillsdale, N.J.: Lawrence Erlbaum Associates.

Brown, R. W. 1973. *A first language.* Cambridge, Mass.: Harvard University Press.

Brown, R. W., and E. H. Lenneberg. 1954. A study in language and cognition. *Journal of Abnormal Social Psychology* 49:454–462.

Bruner, J. S. 1983. *Child's talk.* New York: W. W. Norton.

Bunge, M. 1984. Philosophical problems in linguistics. *Erkenntnis* 21: 107–173.

Caplan, D. 1987. *Neurolinguistics and linguistic aphasiology: An introduction.* Cambridge: Cambridge University Press.

Carew, T. J., E. T. Walters, and E. R. Kandel. 1981. Associative learning in *Aplysia:* Cellular correlates supporting a conditioned fear hypothesis. *Science* 211:501–503.

Carringer, D. 1974. Creative thinking abilities of Mexican youth: The relationship of bilingualism. *Journal of Cross-Cultural Psychology* 5:492–504.

Changeux, J.-P. 1980. Properties of the neuronal network. In *Language and learning: The debate between Jean Piaget and Noam Chomsky,* ed. M. Piatelli-Palmarini, 184–202. Cambridge, Mass.: Harvard University Press.

Cheney, D. L., and R. M. Seyfarth. 1980. Vocal recognition in free-ranging vervet monkeys. *Animal Behavior* 28:362–367.

Chomsky, N. 1957. *Syntactic structures.* The Hague: Mouton.

——— 1959. Review of B. F. Skinner's *Verbal behavior. Language* 3:26–58.

——— 1972. *Language and mind.* Enlarged ed. New York: Harcourt, Brace and World.

——— 1975. *Reflections on language.* New York: Pantheon.

——— 1976. On the nature of language. In *Origins and evolution of language and speech,* ed. H. B. Steklis, S. R. Harnad, and J. Lancaster, 46–57. New York: New York Academy of Sciences.

——— 1980a. Initial states and steady states. In *Language and learning: The debate between Jean Piaget and Noam Chomsky,* ed. M. Piattelli-Palmarini, 107–130. Cambridge, Mass.: Harvard University Press.

——— 1980b. Rules and representations. *Behavioral and Brain Sciences* 3:1–61.

——— 1986. *Knowledge of language: Its nature, origin, and use.* New York: Praeger.

Chomsky, N., and M. Halle. 1968. *The sound pattern of English.* New York: Harper & Row.

Crelin, E. S. 1969. *Anatomy of the newborn: An atlas.* Philadelphia: Lea and Febiger.

Cummings, J. L., and D. F. Benson. 1984. Subcortical dementia: Review of an emerging concept. *Archives of Neurology* 41:874–879.

Dalai Lama, His Holiness the Fourteenth, Tenzin Gyatso. 1984. *Kindness, clarity, and insight,* trans. J. Hopkins. Ithaca, N.Y.: Snow Lion Publications.

D'Antonia, R., J. C. Baron, Y. Samson, M. Serdaru, F. Viader, Y. Agid, and J. Cambier. 1985. Subcortical dementia: Frontal cortex hypometabolism detected by positron tomography in patients with progressive supranuclear palsy. *Brain* 108:785–799.

Darley, F. L., A. A. Aronson, and J. R. Brown. 1975. *Motor speech disorders.* Philadelphia: W. B. Saunders.

Darwin, C. 1859/1964. *On the origin of species.* Facsimile ed. Cambridge, Mass.: Harvard University Press.

——— 1872. *The expression of the emotions in man and animals.* London: John Murray.

Dawkins, R. 1976. *The selfish gene.* New York: Oxford University Press.

Day, M. H. 1986. *Guide to fossil man.* 4th ed. Chicago: University of Chicago Press.

Deacon, T. W. 1984. Connections of the inferior periarcuate area in the brain of *Macaca fascicularis:* An experimental and comparative neuroanatomical investigation of language circuitry and its evolution. Ph.D. diss., Harvard University.

——— 1988a. Human brain evolution I. Evolution of language circuits. In *Intelligence and evolutionary biology,* ed. H. J. Jerison and I. Jerison, 363–382. Berlin: Springer-Verlag.

——— 1988b. Human brain evolution II. Embryology and brain allometry. In *Intelligence and evolutionary biology,* ed. H. J. Jerison and I. Jerison, 383–416. Berlin: Springer-Verlag.

——— 1990. The neural circuitry underlying primate calls and human language. In *The origin of language. Proceedings of a NATO/Advanced Study Institute,* ed. B. A. Chiarelli, P. Lieberman, and J. Wind. Florence: Il Sedicesimo.

DeLong, M. R., A. P. Georgopoulos, and M. D. Crutcher. 1983. Cortico-basal ganglia relations and coding of motor performance. In *Neural coding of motor performance,* ed. J. Massion, J. Paillard, W. Schultz, and M. Wiesendanger, 30–40. Berlin: Springer-Verlag.

Dibble, H. 1989. The implications of stone tool types for the presence of language during the lower and middle Palaeolithic. In *The human revolution: Behavioural and biological perspectives in the origins of modern humans,* ed. P. Mellars and C. B. Stringer, 415–432. Edinburgh: Edinburgh University Press.

Dresher, B. E. 1989. A parameter-based learning model for metrical phonology. In *Conference on language development (1989),* 3. Boston: Program in Applied Linguistics, Boston University.

DuBrul, E. L. 1977. Origins of the speech apparatus and its reconstruction in fossils. *Brain and Language* 4:365–381.

Edelman, G. M. 1987. *Neural Darwinism.* New York: Basic Books.

Eldridge, N., and S. J. Gould. 1972. Punctuated equilibria: An alternative to phyletic gradualism. In *Models in paleobiology,* ed. T. J. M. Schopf. San Francisco: Freeman Cooper.

Engen, E., and T. Engen, 1983. *Rhode Island test of language structure.* Baltimore: University Park Press.

Evarts, E. V. 1973. Motor cortex reflexes associated with learned movement. *Science* 179:501–503.

Exner, S. 1881. *Untersuchungen über Localisation der Functionen in der Grosshirnrinde des Menschen.* Vienna: W. Braumuller.

Falk, D. 1975. Comparative anatomy of the larynx in man and the chimpanzee: Implications for language in Neanderthal. *American Journal of Physical Anthropology* 43:123–132.

Fant, G. 1956. On the predictability of formant levels and spectrum envelopes from formant frequencies. In *For Roman Jakobson,* ed. M. Halle, H. Lunt, and H. MacLean, 104–130. The Hague: Mouton.

——— 1960. *Acoustic theory of speech production.* The Hague: Mouton.

Fernald, A. 1982. Acoustic determinants of infant preference for "motherese." Ph.D. diss., University of Oregon.

Fischer, K. W. 1980. A theory of cognitive development: The control and construction of hierarchies of skills. *Psychological Review* 87:477–531.

Fischer, K. W., and D. Bullock. 1986. Cognitive development in school-age children: Conclusions and new directions. In *Development during middle childhood: The years from six to twelve,* ed. W. A. Collins, 70–146. Washington, D.C.: National Academy of Sciences Press.

Fleming, H. 1988. Mother tongue. *Newsletter of the Association for the Study of Language in Prehistory.*

Flowers, K. A., and C. Robertson. 1985. The effects of Parkinson's disease on the ability to maintain a mental set. *Journal of Neurology, Neurosurgery, and Psychiatry* 48:517–529.

Fodor, J. 1983. *Modularity of mind.* Cambridge, Mass.: MIT Press.

Fouts, R. S., A. D. Hirsch, and D. H. Fouts. 1982. Cultural transmission of a human language in a chimpanzee mother-infant relationship. In *Child nurturance,* ed. H. E. Fitzgerald, J. A. Mullins, and P. Gage. Vol. 3, 159–193. New York: Plenum Press.

Fuster, J. M. 1980. *The prefrontal cortex: Anatomy, physiology, and neuropsychology of the frontal lobe.* New York: Raven Press.

Gall, F. J. 1809. *Recherches sur le système nerveux.* Paris: B. Baillière.

Gardner, H. 1983. *Frames of mind.* New York: Basic Books.

Gardner, R. A., and B. T. Gardner. 1969. Teaching sign language to a chimpanzee. *Science* 165:664–672.

——— 1984. A vocabulary test for chimpanzees (*Pan troglodytes*). *Journal of Comparative Psychology* 4:381–404.

———— 1988. Feedforward vs. feedbackward: An ethological alternative to the law of effect. *Behavioral and Brain Sciences* 11:429–446.

Gazdar, G. 1981. Phrase structure grammar. In *The nature of syntactic representation*, ed. P. Jakobson and G. K. Pullum. Dordrecht: Reidel.

Geschwind, N. 1964. The development of the brain and the evolution of language. *Georgetown Monograph Series on Language and Linguistics* 17:155–169.

———— 1965. Disconnection syndromes in animals and man. Parts I and II. *Brain* 88:237–294, 585–664.

Geschwind, N., and P. O. Behan. 1984. Laterality, hormones, and immunity. In *Cerebral dominance: The biological foundations*, ed. N. Geschwind and A. M. Galaburda, 211–226. Cambridge, Mass.: Harvard University Press.

Gleitman, L. R., H. Gleitman, B. Landau, and E. Wanner. 1987. Where learning begins: Initial representations for language learning. In *The Cambridge Linguistic Survey*, ed. F. Newmeyer, 150–193. New York: Cambridge University Press.

Goldstein, K. 1948. *Language and language disturbances.* New York: Grune and Stratton.

Goodall, J. 1986. *The chimpanzees of Gombe: Patterns of behavior.* Cambridge, Mass.: Harvard University Press.

Gopnick, A., and A. Meltzoff. 1985. From people, to plans, to objects: Changes in the meanings of early words and their relation to cognitive development. *Journal of Pragmatics* 9:495–512.

———— 1986. Words, plans, things, and locations: Interactions between semantic and cognitive development in the one-word stage. In *The development of word meaning*, ed. S. A. Kuczaj and M. D. Barrett, 199–223. New York: Springer-Verlag.

———— 1987. The development of categorization in the second year and its relation to other cognitive and linguistic developments. *Child Development* 58:1523–31.

Gould, S. J., and N. Eldridge. 1977. Punctuated equilibria: The tempo and mode of evolution reconsidered. *Paleobiology* 3:115–151.

Gracco, V., and J. Abbs. 1985. Dynamic control of the perioral system during speech: Kinematic analyses of autogenic and nonautogenic sensorimotor processes. *Journal of Neurophysiology* 54:418–432.

Greenberg, J. 1963. *Universals of language.* Cambridge, Mass.: MIT Press.

Greenberg, S. M., V. A. Castellucci, H. Bayley, and J. H. Schwartz. 1987. A molecular mechanism for long-term sensitization in *Aplysia. Nature* 329:62–65.

Greenewalt, C. A. 1968. *Bird song: Acoustics and physiology.* Washington, D.C.: Smithsonian Institution Press.

Greenfield, P. M., and S. Savage-Rumbaugh. In press. Imitation, grammatical development, and the invention of protogrammar by an ape. In *Biological foundations of language development*, ed. N. Krasnagor, D. M. Rumbaugh, M. Studdert-Kennedy, and D. Scheifel-busch. Hillsdale, N.J.: Lawrence Erlbaum Associates.

Grieser, D. L., and P. K. Kuhl. 1988. Maternal speech to infants in a tonal language: Support for universal prosodic features in motherese. *Developmental Psychology* 24:14–20.

———— 1989. Categorization of speech by infants: Support for speech-sound prototypes. *Developmental Psychology* 25:577–588.

Hamilton, W. D. 1964. The genetical evolution of social behavior, parts 1 and 2. *Journal of Theoretical Biology* 7:1–52.

Hayes, K. J., and C. Hayes. 1951. The intellectual development of a home-raised chimpanzee. *Proceedings of the American Philosophical Society* 95:105–109.

Hebb, D. O. 1949. *The organization of behavior: A neuropsychological theory.* New York: Wiley.

Heffner, R., and H. Heffner. 1980. Hearing in the elephant (*Elephas maximus*). *Science* 208:518–520.

———— 1984. Temporal lobe lesions and perception of species-specific vocalizations by macaques. *Science* 226:75–76.

Heider, E. R. 1972. Universals in color naming and memory. *Journal of Experimental Psychology* 93:10–20.

Herman, L. M., and W. N. Tavolga. 1980. The communication systems of cetaceans. In *Cetacean behavior: Mechanisms and functions.* New York: Wiley.

Herrnstein, R. J. 1979. Acquisition, generalization, and discrimination of a natural concept. *Journal of Experimental Psychology and Animal Behavioral Processes* 5:116–129.

Herrnstein, R. J., and P. A. de Villiers. 1980. Fish as a natural category for people and pigeons. In *The psychology of learning and motivation,* ed. G. H. Bower. Vol. 14, 59–95. New York: Academic Press.

Hewes, G. W. 1973. Primate communication and the gestural origin of language. *Current Anthropology* 14:5–24.

Hirsch-Pasek, K., L. Naigles, R. Golinkoff, L. R. Gleitman, and H. Gleitman. 1988. Syntactic bootstrapping: Evidence from comprehension. In *Proceedings of the 13th Annual Boston University Conference on Language Development.* Vol. 12. Boston: Program in Applied Linguistics, Boston University.

Holloway, R. L. 1985. The poor brain of *Homo sapiens neanderthalensis:* See what you please . . . In *Ancestors: The hard evidence,* ed. E. Delson, 319–324. New York: Alan D. Liss.

Hubel, D. H., and T. N. Wiesel. 1962. Receptive fields, binocular interaction, and functional architecture in the cat's visual cortex. *Journal of Physiology* 160:106–154.

Humboldt, W. von. 1988 (1836). *On language: The diversity of human language-structure and its influence on the mental development of mankind,* trans. P. Heath. Cambridge: Cambridge University Press.

Huxley, J. 1963. *Evolution: The modern synthesis.* New York: Hafner.

Ianco-Worall, A. D. 1972. Bilingualism and cognitive development. *Child Development* 43:1390–1400.

Illes, J., E. J. Metter, W. R. Hanson, and S. Iritani. 1988. Language production in Parkinson's disease: Acoustic and linguistic considerations. *Brain and Language* 33:146–160.

International Phonetic Association. 1949. *The principles of the International Phonetic Association: Being a description of the International Phonetic Alphabet and the manner of using it.* London: Department of Phonetics, University College.

Jakobson, R. 1968 (1940). *Child language, aphasia, and phonological universals,* trans. A. R. Keiler. The Hague: Mouton.

Jerison, H. J. 1973. *Evolution of the brain and intelligence.* New York: Academic Press.

Johnson, J. S., and E. L. Newport, 1989. Critical period effects in second language learning: The influence of maturational state on the acquisition of English as a second language. *Cognitive Psychology* 21:60–99.

Jones, D. 1932. *An outline of English phonetics.* 3d ed. New York: E. P. Dutton.

Kagan, J. 1987. Perspectives on human infancy. In *Handbook of infant development,* ed. J. Osofsky, 1150–98. 2d ed. New York: Wiley.

Kagan, J., J. S. Reznick, and N. Snidman. 1988. Biological bases of childhood shyness. *Science* 240:167–171.

Kant, I. 1981 (1785). *Groundings for the metaphysics of morals,* trans. J. W. Ellington. Indianapolis: Hackett.

Katz, D. I., M. P. Alexander, and A. M. Mandell. 1987. Dementia following strokes in the mesencephalon and diencephalon. *Archives of Neurology* 44:1127–33.

Kempler, D. 1988. Lexical and pantomime abilities in Alzheimer's disease. *Aphasiology* 2:147–159.

Kempler, D., S. Curtiss, and C. Jackson. 1987. Syntactic preservation in Alzheimer's disease. *Journal of Speech and Hearing Research* 30:343–350.

Kimura, D. 1979. Neuromotor mechanisms in the evolution of human communication. In *Neurobiology of social communication in primates,* ed. H. D. Steklis and M. J. Raleigh, 197–219. New York: Academic Press.

——— 1988. Review of H. Poizner, E. S. Klima, and U. Bellugi's *What the hands reveal about the brain. Language and Speech* 31:375–378.

Kimura, D., R. Battison, and B. Lubert. 1976. Impairment of nonlinguistic hand movements in a deaf aphasic. *Brain and Language* 3:566–571.

Kohonen, T. 1984. *Self-organization and associative memory.* New York: Springer-Verlag.

Kruska, D. 1988. Mammalian domestication and its effect on brain structure and behavior. In *Intelligence and evolutionary biology,* ed. H. J. Jerison and I. Jerison, 211–250. Berlin: Springer-Verlag.

Kuhl, P. K. 1988. Auditory perception and the evolution of speech. *Human Evolution* 3:21–45.

Ladefoged, P., and D. E. Broadbent. 1957. Information conveyed by vowels. *Journal of the Acoustical Society of America* 29:98–104.

Ladefoged, P., J. De Clerk, M. Lindau, and G. Papcun. 1972. An auditory-motor theory of speech production. *UCLA Working Papers in Phonetics* 22:48–76.

Laitman, J. T., and E. S. Crelin. 1976. Postnatal development of the basicranium and vocal tract region in man. In *Symposium on development of the basicranium,* ed. J. Bosma, 206–219. Washington, D.C.: U.S. Government Printing Office.

Laitman, J. T., and R. C. Heimbuch. 1982. The basicranium of Plio-Pleistocene hominids as an indicator of their upper respiratory systems. *American Journal of Physical Anthropology* 59:323–344.

Laitman, J. T., R. C. Heimbuch, and E. S. Crelin. 1978. Developmental change in a basicranial line and its relationship to the upper respiratory system in living primates. *American Journal of Anatomy* 152:467–482.

—— 1979. The basicranium of fossil hominids as an indicator of their upper respiratory systems. *American Journal of Physical Anthropology* 51:15–34.

Laitman, J. T., and J. S. Reidenberg. 1988. Advances in understanding the relationship between the skull base and larynx, with comments on the origins of speech. *Human Evolution* 3:101–111.

Laitman, J. T., J. S. Reidenberg, P. J. Gannon, B. Johansson, K. Landahl, and P. Lieberman. 1990. The Kebara hyoid: What can it tell us about the evolution of the hominid vocal tract? *American Journal of Physical Anthropology* 81:254.

Landahl, K. L., and H. J. Gould. 1986. Congenital malformation of the speech tract in humans and its developmental consequences. In *The biology of change in otolaryngology,* ed. R. J. Ruben, T. R. Van de Water, and E. W. Rubel, 131–149. Amsterdam: Elsevier.

Landau, B., and L. R. Gleitman. 1985. *Language and experience: Evidence from the blind child.* Cambridge, Mass.: Harvard University Press.

Lenneberg, E. H. 1967. *Biological foundations of language.* New York: Wiley.

Liberman, A. M., F. S. Cooper, D. P. Shankweiler, and M. Studdert-Kennedy. 1967. Perception of the speech code. *Psychological Review* 74:431–461.

Liberman, A. M., and I. G. Mattingly. 1985. The motor theory of speech perception revised. *Cognition* 21:1–36.

Liberman, F. Z. 1979. Learning by neural nets. Ph.D. diss., Brown University.

Lieberman, M. R., and P. Lieberman. 1973. Olson's "projective verse" and the use of breath control as a structural element. *Language and Style* 5:287–298.

Lieberman, P. 1967. *Intonation, perception, and language.* Cambridge, Mass.: MIT Press.

——— 1968. Primate vocalizations and human linguistic ability. *Journal of the Acoustical Society of America* 44:1157–64.

——— 1975. *On the origins of language: An introduction to the evolution of speech.* New York: Macmillan.

——— 1984. *The biology and evolution of language.* Cambridge, Mass.: Harvard University Press.

——— 1985. On the evolution of human syntactic ability: Its pre-adaptive bases—motor control and speech. *Journal of Human Evolution* 14: 657–668.

——— 1989. The origins of some aspects of human language and cognition. In *The human revolution: Behavioural and biological perspectives in the origins of modern humans,* ed. P. Mellars and C. B. Stringer, 391–414. Edinburgh: Edinburgh University Press.

Lieberman, P., and S. E. Blumstein. 1988. *Speech physiology, speech perception, and acoustic phonetics.* Cambridge: Cambridge University Press.

Lieberman, P., and E. S. Crelin. 1971. On the speech of Neanderthal man. *Linguistic Inquiry* 2:203–222.

Lieberman, P., E. S. Crelin, and D. H. Klatt. 1972. Phonetic ability and related anatomy of the newborn, adult human, Neanderthal man, and the chimpanzee. *American Anthropologist* 74:287–307.

Lieberman, P., L. S. Feldman, S. Aronson, and E. Engen. 1989. Sentence comprehension, syntax, and vowel duration in aged people. *Clinical Linguistics and Phonetics* 3:299–311.

Lieberman, P., J. Friedman, and L. S. Feldman. 1990. Syntactic deficits in Parkinson's disease. *Journal of Nervous and Mental Disease* 178: 360–365.

Lieberman, P., J. Friedman, G. Tajchman, L. S. Feldman, and E. Kako. In preparation. Broca's aphasia-like speech and syntax deficits in Parkinson's disease: A voice onset time study.

Lieberman, P., K. S. Harris, P. Wolff, and L. H. Russell. 1972. Newborn infant cry and nonhuman primate vocalizations. *Journal of Speech and Hearing Research* 14:718–727.

Lieberman, P., D. H. Klatt, and W. H. Wilson. 1969. Vocal tract limitations on the vowel repertoires of rhesus monkey and other nonhuman primates. *Science* 164:1185–87.

Lieberman, P., J. T. Laitman, J. S. Reidenberg, K. Landahl, and P. J. Gannon. 1989. Folk physiology and talking hyoids. *Nature* 342:486–487.

Lieberman, P., R. H. Meskill, M. Chatillon, and H. Schupack. 1985. Phonetic speech deficits in dyslexia. *Journal of Speech and Hearing Research* 28:480–486.

Liepmann, H. 1908. *Drei Aufsatze aus dem Apraxiegebiet.* Berlin: Karger.

Lindblom, B. 1988. Models of phonetic variation and selection. *Language change and biological evolution.* Turin: Institute for Scientific Interchange.

Linebarger, M., M. Schwartz, and E. Saffran. 1983. Sensitivity to grammatical structure in so-called agrammatic aphasics. *Cognition* 13: 361–392.

Long, C. A. 1969. The origin and evolution of mammary glands. *Biological Sciences* 19:519–523.

Lorenz, K. 1974. Analogy as a source of knowledge. *Science* 185:229–234.

Lubker, J., and T. Gay. 1982. Anticipatory labial coarticulation: Experimental, biological, and linguistic variables. *Journal of the Acoustical Society of America* 71:437–438.

Lupker, S. 1984. Semantic priming without association: A second look. *Journal of Learning and Verbal Behavior* 23:709–733.

Luria, A. R. 1973. The frontal lobes and the regulation of behavior. In *Psychobiology of the frontal lobes,* ed. K. H. Pribram and A. R. Luria, 3–26. New York: Academic Press.

MacLean, P. D. 1967. The brain in relation to empathy and medical education. *Journal of Nervous and Mental Disorders* 144:374–382.

——— 1973. A triune concept of the brain and behavior. In *The Hincks Memorial Lectures,* ed. T. Boag and D. Campbell, 6–66. Toronto: University of Toronto Press.

——— 1985. Evolutionary psychiatry and the triune brain. *Psychological Medicine* 15:219–221.

——— 1986. Neurobehavioral significance of the mammal-like reptiles (therapsids). In *The ecology and biology of mammal-like reptiles,* ed. N. Hotton III, J. J. Roth, and E. C. Roth, 1–21. Washington, D.C.: Smithsonian Institution Press.

MacLean, P. D., and J. D. Newman. 1988. Role of midline frontolimbic cortex in the production of the isolation call of squirrel monkeys. *Brain Research* 450:111–123.

MacNeilage, P. F. 1987. The evolution of hemispheric specialization for manual function and language. In *Higher brain functions: Recent explorations of the brain's emergent properties,* ed. S. P. Wise. New York: Wiley.

MacNeilage, P. F., M. G. Studdert-Kennedy, and B. Lindblom. 1987. Primate handedness reconsidered. *Behavioral and Brain Sciences* 10: 247–303.

MacNeill, D. 1985. So you think gestures are nonverbal? *Psychological Review* 92:350–371.

Manley, R. S., and L. C. Braley. 1950. Masticatory performance and efficiency. *Journal of Dental Research* 29:314–321.

Manley, R. S., and F. R. Shiere. 1950. The effect of dental efficiency on mastication and food preference. *Oral Surgery, Oral Medicine, and Oral Pathology* 3:674–685.

Markowitsch, H. J. 1988. Anatomical and functional organization of the primate prefrontal cortical system. In *Comparative primate biology,* vol. 4: *Neurosciences,* ed. H. D. Steklis and J. Erwin, 99–153. New York: Alan D. Liss.

Marler, P. 1976. An ethological theory of the origin of vocal learning. In *Origins and evolution of language and speech,* ed. S. R. Harnad, H. D. Steklis, and J. Lancaster, 386–395. New York: New York Academy of Science.

Marler, P., and M. Tamura. 1964. Culturally transmitted patterns of vocal behavior in sparrows. *Science* 146:1483–86.

Marshack, A. 1990. The origin of language: An anthropological approach. In *The origin of language: Proceedings of a NATO/Advanced Study Institute,* ed. B. A. Chiarelli, P. Lieberman, and J. Wind. Florence: Il Sedicesimo.

Maynard-Smith, J. 1978. The evolution of behavior. *Scientific American* 239:176–192.

Mayr, E. 1982. *The growth of biological thought.* Cambridge, Mass.: Harvard University Press.

McCarthy, R. A., and E. K. Warrington. 1988. Evidence for modality-specific meaning systems in the brain. *Nature* 334:428–430.

McCawley, J. D. 1988. Comments on Noam A. Chomsky, "Language and problems of knowledge" (MIT Cognitive Science Colloquium, March 11, 1987). *University of Chicago Working Papers in Linguistics* 4:148–156.

McCowan, T. D., and A. Keith. 1939. *The stone age of Mount Carmel. Vol. 2: The fossil human remains from the Levalloisio-Mousterian.* Oxford: Clarendon Press.

Meltzoff, A. N. 1988. Imitation, objects, tools, and the rudiments of language in human ontogeny. *Human Evolution* 3:47–66.

Meltzoff, A. N., and M. K. Moore. 1977. Imitation of facial and manual gestures by human neonates. *Science* 198:75–78.

———— 1983. Newborn infants imitate adult facial gestures. *Child Development* 54:702–709.

Menzel, E. W., Jr. 1978. Cognitive mapping in chimpanzees. In *Cognitive processes in animal behavior,* ed. S. H. Hulse, H. Fowler, and W. K. Honig, 375–422. Hillsdale, N.J.: Lawrence Erlbaum Associates.

Menzel, E. W., Jr., D. Premack, and G. Woodruff. 1978. Map reading by chimpanzees. *Folia Primatologica* 29:241–249.

Merzenich, M. M. 1987a. On the plasticity of cortical maps. In *The neural and molecular bases of learning,* ed. J.-P. Changeux and M. Konishi, 337–358. Chichester: Wiley.

———— 1987b. Cerebral cortex: A quiet revolution in thinking. *Nature* 328:572–573.

Mesulam, M. M. 1985. Patterns in behavioral neuroanatomy: Association areas, the limbic system, and hemispheric specialization. In *Principles of behavioral neurology*, 1–70. Philadelphia: F. A. Davis.

Metter, E. J., D. Kempler, C. A. Jackson, W. R. Hanson, J. C. Mazziotta, and M. E. Phelps, 1989. Cerebral glucose metabolism in Wernicke's, Broca's, and conduction aphasia. *Archives of Neurology* 46:27–34.

Metter, E. J., D. Kempler, C. A. Jackson, W. R. Hanson, W. H. Reige, L. M. Camras, J. C. Mazziotta, and M. E. Phelps. 1987. Cerebular glucose metabolism in chronic aphasia. *Neurology* 37:1599–1606.

Metter, E. J., W. H. Reige, W. R. Hanson, M. E. Phelps, and D. E. Kuhl. 1984. Local cerebral metabolic rates of glucose in movement and language disorders from positron tomography. *American Journal of Physiology* 246:R897–900.

Milberg, W., S. E. Blumstein, and B. Dworetzky. 1985. Sensitivity to morphological constraints in Broca's and Wernicke's aphasics: A double dissociation of syntactic judgments and syntactic facilitations in a lexical decision task. Paper presented at the annual meeting of the Academy of Aphasia, Pittsburg.

Miles, F. A., and E. V. Evarts. 1979. Concepts of motor organization. *Annual Reviews of Psychology* 30:327–362.

Miller, G. A. 1956. The magical number seven, plus or minus two: Some limits on our capacity for processing information. *Psychological Review* 63:81–97.

Miller, G. A., and P. E. Nicely. 1955. An analysis of perceptual confusions among some English consonants. *Journal of the Acoustical Society of America* 27:338–352.

Milner, B. 1964. Some effects of frontal lobectomy in man. In *The frontal granular cortex and behavior*, ed. J. M. Warren and K. Akert, 313–334. New York: McGraw-Hill.

Mohanty, A. K., and K. Pattwaik. 1984. Relationship between metalinguistics and cognitive development of bilingual and unilingual tribal children. *Psycho-Lingua* 14:63–70.

Müller, J. 1848. *The physiology of the senses, voice and muscular motion with the mental faculties*, trans. W. Baly. London: Walton and Maberly.

Naeser, M. A., M. P. Alexander, N. Helms-Estabrooks, H. L. Levine, S. A. Laughlin, and N. Geschwind. 1982. Aphasia with predominantly subcortical lesion sites: Description of three capsular/putaminal aphasia syndromes. *Archives of Neurology* 39:2–14.

Nearey, T. 1978. *Phonetic features for vowels*. Bloomington: Indiana University Linguistics Club.

Negus, V. E. 1949. *The comparative anatomy and physiology of the larynx*. New York: Hafner.

Nelson, K. 1975. The nominal shift in semantic-syntactic development. *Cognitive Psychology* 7:461–479.

Newman, J. D. 1985. The infant cry of primates: An evolutionary perspective. In *Infant crying*, ed. B. M. Lester and C. F. Zachariah Boukydis, 307–324. New York: Plenum Press.

———— 1988. Primate hearing mechanisms. In *Comparative primate biology*, vol. 4: *Neurosciences*, ed. H. D. Steklis and J. Erwin, 469–499. New York: Alan R. Liss.

———— 1990. The primate isolation call and the evolution and physiological control of human speech. In *The origin of language: Proceedings of a NATO/Advanced Study Institute*, ed. B. A. Chiarelli, P. Lieberman, and J. Wind. Florence: Il Sedicesimo.

Newman, J. D., and P. D. MacLean. 1982. Effects of tegmental lesions on the isolation call of squirrel monkeys. *Brain Research* 232:317–329.

Newport, E. L., L. R. Gleitman, and H. R. Gleitman. 1977. Mother, I'd rather do it myself: Some effects and non-effects of maternal speech style. In *Talking to children: Language input and acquisition*, ed. C. E. Snow and C. A. Ferguson. Cambridge: Cambridge University Press.

Newport, E. L., and T. Suppala. 1987. A critical period effect in the acquisition of a primary language. University of Illinois.

Nicholas, M., L. Obler, M. Albert, and J. Helm-Estabrooks. 1985. Empty speech in Alzheimer's disease and fluent aphasia. *Journal of Speech and Hearing Research* 28:405–410.

North, G. 1987. Neural networks: Implementation and analysis. *Nature* 330:522–523.

Nottebohm, F. 1984. Vocal learning and its possible relation to replaceable synapses and neurons. In *Biological perspectives on language*, ed. D. Caplan. Cambridge, Mass.: MIT Press.

Ojemann, G. A. 1983. Brain organization for language from the perspective of electrical stimulation mapping. *Behavioral and Brain Sciences* 2:189–230.

Okoh, N. 1980. Bilingualism and divergent thinking among Nigerian and Welsh school children. *Journal of Social Psychology* 10:163–170.

Olmsted, D. L. 1971. *Out of the mouths of babes*. The Hague: Mouton.

Parent, A. 1986. *Comparative neurobiology of the basal ganglia*. New York: Wiley.

Parker, G. A. 1978. Searching for mates. In *Behavioral ecology*, ed. J. R. Krebs and N. B. Davies. Oxford: Blackwell Scientific.

Peal, E., and W. E. Lambert. 1962. The relationship of bilingualism to intelligence. *Psychological Monographs: General and Applied* 76:1–23.

Pepperberg, I. M. 1981. Functional vocalizations by an African grey parrot (*Psittacus erithacus*). *Zeitschrift für Tierpsychologie* 55:139–160.

Perkell, J. S. 1969. *Physiology of speech production: Results and implications of a quantitative cineradiographic study*. Cambridge, Mass.: MIT Press.

Peterson, G. E., and H. L. Barney, 1952. Control methods used in a study of the vowels. *Journal of the Acoustical Society of America* 24:175–184.

Peterson, M. R., M. D. Deecher, S. R. Zolith, D. B. Moody, and W. C. Stebbens. 1978. Species-specific perceptual processing of vocal sounds by monkeys. *Science* 202:324–326.

Pfungst, O. 1907. *Das Pfred des Herrn von Osten (Der kluge Hans).* Leipzig: J. Ambrosius.

Piaget, J. 1952. *The origins of intelligence in children,* trans. M. Cook. New York: W. W. Norton.

———— 1962. *Play, dreams, and imitation in childhood,* trans. G. Gattegno and F. M. Hodgson. New York: W. W. Norton.

Pierce, J. D. 1985. A review of attempts to condition operantly Alloprimate vocalizations. *Primates* 26:202–213.

Pillon, P., B. Dubois, F. Lhermitte, and Y. Agid. 1986. Heterogeneity of cognitive impairment in progressive supranuclear palsy, Parkinson's disease, and Alzheimer's disease. *Neurology* 36:1179–85.

Pinker, S. 1984. *Language learnability and language development.* Cambridge, Mass.: Harvard University Press.

Pirozzolo, F. J., E. C. Hansch, J. A. Mortimer, D. D. Webster, and M. A. Kuskowski. 1982. Dementia in Parkinson's disease: A neuropsychological analysis. *Brain and Cognition* 1:71–83.

Poizner, H., E. S. Klima, and U. Bellugi. 1987. *What the hands reveal about the brain.* Cambridge, Mass.: MIT Press.

Polit, A., and E. Bizzi. 1978. Processes controlling arm movements in monkeys. *Science* 201:1235–37.

Posner, M. I., and C. Snyder. 1975. Attention and cognitive control. In *Information processing and cognition,* ed. R. Solso. Hillsdale, N.J.: Lawrence Erlbaum Associates.

Posner, M. I., S. E. Petersen, P. T. Fox, and M. E. Raichle. 1988. Localization of cognitive functions in the human brain. *Science* 240:1627–31.

Powers, S., and R. L. Lopez. 1985. Perceptual, motor, and verbal skills of monolingual and bilingual Hispanic children: A discriminant analysis. *Perceptual and Motor Skills* 60:999–1002.

Premack, D. 1988. Lecture at Brown University.

Premack, D., and G. Woodruff. 1978. Does the chimpanzee have a theory of mind? *Brain and Behavior Sciences* 1:515–526.

Pruzansky, S. 1973. Clinical investigations of the experiments of nature. In *Orofacial anomalies: Clinical and research implications,* 62–94. Washington, D.C.: American Speech and Hearing Association.

Putnam, H. 1981. *Reason, truth, and history.* Cambridge: Cambridge University Press.

Richman, B. 1976. Some vocal distinctive features used by gelada monkeys. *Journal of the Acoustical Society of America* 60:718–724.

Rumelhart, D. E., J. L. McClelland, and the PDP Research Group. 1986.

Parallel distributed processing: Explorations in the microstructures of cognition. Cambridge, Mass.: MIT Press.

Rushton, J. P., D. W. Fulker, M. C. Neale, D. K. B. Nias, and H. J. Eysenck. 1985. Altruism and aggression: Individual differences are substantially inheritable. *Journal of Personality and Social Psychology* 41:459–466.

Russell, G. O. 1928. *The vowel.* Columbus: Ohio State University Press.

Sapir, E. 1949. *Selected writings of Edward Sapir in language, culture, and personality,* ed. D. G. Mandelbaum. Berkeley: University of California Press.

Saravia, A. E. 1977. *Popul Wuh: Ancient Stories of the Quiche Indians of Guatemala.* Guatemala City: Publicaciones Turisticas.

Sarich, V. M. 1974. Just how old is the hominid line? In *Yearbook of physical anthropology, 1973.* Washington, D.C.: American Association of Physical Anthropologists.

Savage-Rumbaugh, S., K. McDonald, R. A. Sevcik, W. D. Hopkins, and E. Rubert. 1986. Spontaneous symbol acquisition and communicative use by pygmy chimpanzees (*Pan paniscus*). *Journal of Experimental Psychology: General* 115:211–235.

Savage-Rumbaugh, S., D. Rumbaugh, and K. McDonald. 1985. Language learning in two species of apes. *Neuroscience and Behavioral Reviews* 9:653–665.

Schieffelin, B. B. 1982. *How Kaluli children learn what to say, what to do, and how to feel: An ethnographic study of the development of communicative competence.* New York: Cambridge University Press.

Schrier, S. 1977. *Abduction algorithms for grammar discovery.* Providence: Division of Applied Mathematics, Brown University.

Schusterman, R. J., and R. Gisiner. 1988. Animal language research: Marine mammals re-enter the controversy. In *Intelligence and evolutionary biology,* ed. H. J. Jerison and I. Jerison, 319–350. Berlin: Springer-Verlag.

Schusterman, R. J., and K. Krieger. 1984. California sea lions are capable of semantic comprehension. *Psychological Record* 34:3–23.

Sejnowski, T. J., C. Koch, and P. S. Churchland. 1988. Computational neuroscience. *Nature* 241:1299–1306.

Sereno, J., S. R. Baum, G. C. Marean, and P. Lieberman. 1987. Acoustic analyses and perceptual data on anticipatory labial coarticulation in adults and children. *Journal of the Acoustical Society of America* 81:512–519.

Sereno, J., and P. Lieberman. 1987. Developmental aspects of lingual coarticulation. *Journal of Phonetics* 15:247–257.

Shallice, T. 1978. The dominant action system: An information processing approach to consciousness. In *The stream of consciousness,* ed. K. S. Pope and J. L. Singer, 117–157. New York: Plenum Press.

Singleton, J. L., and E. L. Newport. 1989. When learners surpass their models: The acquisition of American Sign Language from impoverished input. In *Proceedings of the 14th Annual Boston University Conference on Language Development.* Vol. 15. Boston: Program in Applied Linguistics, Boston University.

Skinner, B. F. 1957. *Verbal behavior.* New York: Appleton-Century-Crofts.

Slotnick, B. M. 1967. Disturbances of maternal behavior in the rat following lesions of the cingulate cortex. *Behavior* 24:204–236.

Snow, C. E. 1977. Mothers' speech research: From input to interaction. In *Talking to children: Language input and acquisition,* ed. C. E. Snow and C. A. Ferguson, 31–49. Cambridge: Cambridge University Press.

Solecki, R. S. 1971. *Shanidar, the first flower people.* New York: Knopf.

Spearman, C. 1904. "General intelligence," objectively determined and measured. *American Journal of Psychology* 15:201–293.

Spurzheim, J. K. 1815. *The physiognomical system of Gall and Spurzheim.* London.

Stamm, J. S. 1955. The function of the medial cerebral cortex in maternal behavior of rats. *Journal of Comparative Physiology and Psychology* 48:347–356.

Stern, J. T., and R. L. Susman. 1983. The locomotor activity of *Australopithecus afarensis. American Journal of Physical Anthropology* 60:279–317.

Sternberg, Robert J. 1985. *Beyond IQ: A triarchic theory of human intelligence.* New York: Cambridge University Press.

Stevens, K. N. 1972. Quantal nature of speech. In *Human communication: A unified view,* ed. E. E. David, Jr., and P. B. Denes, 51–66. New York: McGraw-Hill.

Stringer, C. B., and P. Andrews. 1988. Genetic and fossil evidence for the origin of modern humans. *Science* 239:1263–68.

Stuss, D. T., and D. F. Benson. 1986. *The frontal lobes.* New York: Raven.

Sutton, D., and U. Jurgens. 1988. Neural control of vocalization. In *Comparative primate biology,* vol. 4: *Neurosciences,* ed. H. D. Steklis and J. Erwin, 625–647. New York: Alan D. Liss.

Taylor, A. E., J. A. Saint-Cyr, and A. E. Lang. 1986. Frontal lobe dysfunction in Parkinson's disease. *Brain* 109:845–883.

Terrace, H. S., L. A. Petitto, R. J. Sanders, and T. G. Bever. 1979. Can an ape create a sentence? *Science* 206:821–901.

Teuber, H. L. 1964. The riddle of frontal lobe function in man. In *The frontal granular cortex and behavior,* ed. J. M. Warren and K. Akert, 410–444. New York: McGraw-Hill.

Thorndike, E. L. 1913. *Educational psychology: The psychology of learning.* New York: Teachers College.

Tomasello, M., M. Davis-Dasilva, L. Camak, and K. Bard. 1987. Observational learning of tool use by young chimpanzees. *Human Evolution* 2:175–183.

Tomasello, M., and M. J. Farrar. 1986. Joint attention and early language. *Child Development* 57:1454–63.

Tomasello, M., S. Mannle, and A. C. Kruger. 1986. Linguistic environment of 1-to-2-year-old twins. *Developmental Psychology* 22:169–176.

Trinkaus, E., and W. W. Howells. 1979. The Neanderthals. *Scientific American* 241:118–133.

Tweedy, J. P., K. G. Langer, and F. A. McDowell. 1982. The effect of semantic relations on the memory deficit associated with Parkinson's disease. *Journal of Clinical Neuropsychology* 4:235–247.

Tyler, L. 1985. Real-time comprehension processes in agrammatism: A case study. *Brain and Language* 26:259–275.

———— 1986. Spoken language in a fluent aphasic. Manuscript Department of Psychology, Cambridge University.

Udwin, O., W. Yule, and N. Martin. 1986. Cognitive abilities and behavioral characteristics of children with idiopathic infantile hypercacaemia. *Child Psychology and Psychiatry* 28:297–309.

Valladas, H., J. L. Joron, G. Valladas, B. Arensburg, O. Bar-Yosef, A. Belfer-Cohen, P. Goldberg, H. Laville, L. Meignen, Y. Rak, E. Tchernov, A. M. Tiller, and B. Vandermeersch. 1987. Thermoluminescence dates for the Neanderthal burial site at Kebara in Israel. *Nature* 330:159–160.

Van Cantfort, T. E., and J. B. Rimpau. 1982. Sign language studies with children and chimpanzees. *Sign Language Studies* 34:15–72.

Vandermeersch, B. 1981. *Les hommes fossiles de Qafzeh, Israel.* Paris: CNRS.

Walters, T., T. J. Carew, and E. R. Kandel. 1981. Associative learning in *Aplysia*: Evidence for conditioned fear in an invertebrate. *Science* 211:404–506.

Warden, C. J., and L. H. Warner. 1928. The sensory capacities and intelligence of dogs, with a report on the ability of the noted dog "Fellow" to respond to verbal stimuli. *Quarterly Review of Biology* 3:1–28.

Washburn, S. L. 1961. *Social life of early man.* Chicago: Aldine.

———— 1969. The evolution of human behavior. In *The uniqueness of man,* ed. J. D. Roslansky. Groningen: North Holland.

Watkins, K., and D. Fromm. 1984. Labial coordination in children: Preliminary considerations. *Journal of the Acoustical Society of America* 75:629–632.

Waxman, S. R. 1985. Hierarchies in classification and language: Evidence from preschool children. Ph.D. diss. University of Pennsylvania.

Wechsler, D. 1944. *Measurement of adult intelligence.* Baltimore: Williams and Wilkins.

Wernicke, C. 1965 (1874). The aphasic symptom complex: A psychological study on a neurological basis. Reprinted in *Boston studies in the philosophy of science*, ed. R. S. Cohen and M. W. Wartofsky. Vol. 4. Boston: Reidel.

Westoll, T. S. 1945. The mammalian middle ear. *Nature* 155:114–115.

White, R. 1987. Body ornamentation in the Upper Paleolithic: The origin of the modern human mind. *Science* 236:670.

Whorf, B. L. 1956. *Language, thought, and reality: Selected writings of Benjamin Lee Whorf*, ed. J. B. Carroll. Cambridge, Mass.: MIT Press.

Wills, R. H. 1973. *The institutionalized severely retarded*. Springfield, Ill.: Charles C. Thomas.

Wilson, E. O. 1975. *Sociobiology: The new synthesis*. Cambridge, Mass.: Harvard University Press.

Wulfeck, B. B. 1988. Grammaticality judgments and sentence comprehension in agrammatic aphasia. *Journal of Speech and Hearing Research* 31:72–81.

Zahn-Wexler, C., E. M. Cummings, D. H. McKnew, and M. Radke-Yarrow. 1984. Altruism, aggression, and social interactions in young children with a manic-depressive parent. *Child Development* 55:112–122.

Zahn-Waxler, C., B. Hollenbeck, and M. Radke-Yarrow. 1984. The origins of empathy and altruism. In *Advances in animal welfare science, 1984/85*, ed. M. W. Fox and L. D. Mickley, 21–39. Washington, D.C.: Humane Society of the United States.

Zentall, T. R., and B. G. Galef. 1988. *Social learning: A comparative approach*. Hillsdale, N.J.: Lawrence Erlbaum Associates.

Zurif, E. B., and S. E. Blumstein. 1978. Language and the brain. In *Linguistic theory and psychological reality*, ed. M. Kalle, J. Bresnan, and G. A. Miller. Cambridge, Mass.: MIT Press.

Zurif, E. B., and A. Caramazza. 1976. Psycholinguistic structures in aphasia: Studies in syntax and semantics. In *Studies in neurolinguistics*, ed. H. Whittaker and H. A. Whittaker. Vol. 1. New York: Academic Press.

Zurif, E. B., A. Caramazza, and R. Meyerson. 1972. Grammatical judgments of agrammatic aphasics. *Neuropsychologia* 10:405–418.

Index

Abbs, James, 50
Acoustic cues and signals, 86, 90, 179n14, 180n1, 181n5
Acoustic energy, 41, 42, 44
Acoustic salience, 57–58, 59
Acoustic stability, 58, 59
Agrammatism, 87–89, 91, 177n6, 178n10
Albert, Martin, 95
Alexander, Michael, 93, 179n13
Altruism: animal, 152, 164–165, 169–171; selfless behavior, 164–171; higher, 165–169; cognitive, 169–171
Alzheimer's disease, 119–120, 121, 178n9
American Sign Language (ASL), 82, 90; chimpanzees and, 113, 155–156; language acquisition in, 136–137; syntax in, 136, 157; Universal Grammar and, 136–137; error rates in, 137–138; aphasia and, 178n7
Anderson, Alun, 34–35, 117
Animals (general discussion): evolution of, 2; word use by, 112; thought process in, 125; learning in, 139–140; altruism in, 152, 164–165
Anticipatory coarticulation, 50
Apert's syndrome, 60, 61, 62, 63
Apes: language ability, 1, 82, 112; brain size, 25, 101; vocalization and vocal tracts, 52, 62, 72, 84–85. *See also* Monkeys
Aphasia, 15, 104; Broca's, 27, 34, 84–90, 93, 94, 97, 98, 99, 177n4; conduction,

27, 94; Wernicke's, 27, 34, 90, 91, 94, 121–122, 178n8; agrammatic, 87–89, 91, 177n6, 178n10; subcortical brain structures and, 90–99; cognitive deficits of, 91–92; global, 93; ASL and, 178n7
Aphemia, 177n5
Apraxia, 177n5
Arensburg, Baruch, 69
Art of early hominids, 161
ASL. *See* American Sign Language
Associative learning, 31, 127, 135, 138–140
Au, Terry, 143–144
Auditory systems, 3, 45–46, 82
Australopithecines, 72, 74, 80, 105, 176nn1,7
Automatization of speech production, 48–51

Barney, Harold, 61, 62
Basal ganglia, 16–17, 21, 22, 25, 52–53, 88, 100, 101–103, 178n11; damage and disease, 85, 93, 94, 99; dopamine and, 95–96, size and structure, 101–102; speech control and, 107
Baum, Shari, 90, 177n6
Bavin, E. C., 177n2
Behavior: changes, 7, 8–10, 18–20; deficits, 27; brain circuitry and, 28–30, 78; aphasia and, 94; prefrontal cortex and, 100
Bellugi, Ursula, 90, 178nn7,10

Benson, D. Frank, 85, 87, 98, 99–100, 177n5
Ben-Zeev, Sandra, 147
Bilingualism, 146–148
Birds, 34, 59, 112, 140, 180n1, 181n4
Bloom, Alfred, 143
Boehm, Christopher, 152
Boesch, Christophe, 151
Boesch, Hedwige, 151
Bon-Po shamanism, 171
Boule, Marcellin, 65
Bowerman, Melissa, 135
Brain and brain mechanisms: speech production and, 1, 3–4, 17, 20, 24, 36, 45–46, 72, 74, 79, 82–83, 85–86, 106–107; neural networks, 3, 11, 17; cognition and, 4; evolution of, 9, 11, 15, 20–21, 78, 80–81, 84, 95, 101, 109–110, 175n3; modular theory of, 12–14, 16; connectionist model of, 15, 27; language and, 15–16, 36; comparative neurophysiological model, 16–27; specialized, 20; locationist model, 25, 28; dictionary, 35, 118–119, 120, 121–126; hemispheres, 79, 84, 90, 91, 105, 118; lateralization, 79–80, 85, 104–105, 149; damage, 92–99, 118, 149, 177n4; biological fitness model, 105; differences, 118–119. *See also* Aphasia; Circuitry model of the brain; Distributed neural networks
Brain structure: components, 16–17; size, 20–21, 22, 25, 26, 29, 63, 101, 110, 175n1, 176n3; cytoarchitectonic, 24–25, 26, 175n1. *See also specific structures and areas*
Breath-groups, 3, 42, 146, 176n1
Breathing and speech production, 39–40
Broadmann, Korbinian, 24, 25, 28, 101
Broca, Paul, 15, 24, 27, 28
Broca's aphasia, 27, 34, 84–90, 93, 94, 177n4; speech and comprehension deficits and, 97, 98; brain activity in, 99
Broca's area, 23, 24, 25, 27, 104, 111; damage to, 34, 84, 85, 91, 93, 94, 98; motor response patterns and, 51;

speech production and, 106, 107
Broken Hill fossil, 76, 176n4
Bronowski, Jacob, 115
Brown, Jason, 178n11
Brown, Roger, 143
Buddhist doctrine, 166–168
Burial rituals, 162–164, 171

Cats, 29, 33–34, 35, 117
Caudate nucleus, 22, 101, 103
Changeux, Jean-Pierre, 28
Chimpanzees: language abilities, 1, 52, 154–157; word use, 1, 112, 113; anatomical structure, 21; vocalization, 51–53, 72, 74, 155; culture, 66, 150–158, 176n1; skulls, 70, 71; formant frequencies in, 77; speech perception, 77; tool use, 104, 105, 142, 150–151, 176n1; ASL and, 113, 155–156; learning abilities, 142; brain size, 150; cognitive abilities, 157, 158, 161–162; evolution of, 182n1. *See also* Apes
Chinchillas, 46
Chomsky, Noam, 12, 78, 109, 120; Universal Grammar theory, 127–134, 177n2; critique of Skinner, 138–139, 140. *See also* Universal Grammar (UG)
Churchland, Patricia, 116, 180n2
Circuitry model of the brain, 11, 14–16, 78, 81, 178n11; cortical organization and, 22–27, 103; language deficits and, 27; behavior and, 28–30; vision-stabilizing, 29–30; distributed neural networks and, 30; ascending, 93; speech production and, 107
Classification system of language acquisition, 156
Cognition: brain mechanisms and, 4, 12, 24, 34–35, 81, 84, 125; linguistic abilities and, 10; deficits, 17, 27, 87, 96, 98–99, 119; distributed neural networks and, 34–35; information processing and, 37–38; evolution of, 72; aphasia and, 91–92; syntax and, 146; chimpanzee, 157, 158; tool technology as index of, 160–161; altruism and,

168, 169–171. *See also* Language acquisition
Communication: speech evolution and, 1–2; mother-child, 3, 18, 20, 134–135; speed, 3, 59, 77, 80–83; brain structure and, 22; vocal, 57, 58, 77
Comparative Anatomy and Physiology of the Larynx, The (Negus), 64
Comparative neurophysiology, 16–17, 18–27
Comprehension deficits, 96–98. *See also* Speech comprehension deficits
Computed-tomographic (CT) scans, 86–87, 92, 93–94, 99
Computers: -brain analogy, 13–14; distributed neural network simulations, 30, 33–35, 116, 118, 123, 125, 175n5; formant frequency calculations, 44; vocal tract models, 51, 58, 60, 62; word meanings and, 114
Concept formation in language acquisition, 136–137
Conceptualization of tool technology, 158–159
Conditioned reflexes, 125
Connectionist model of brain mechanism, 15, 27
Consonants: stop, 45–46, 50–51, 58, 86, 97; labial, 50–51, 58; velar, 58, 59, 65; nasal, 86
Context-dependent rules, 83
Core-and-flake toolmaking technology, 159–160
Cortex: neocortex, 15, 17, 21, 22, 25–27, 53, 85, 95, 101, 106, 178n11; cingulate, 17, 18, 21, 22–23, 52, 103, 107; functional organization of, 22–27, 117–118; prefrontal, 23, 24, 26, 51, 85, 94, 99–101, 103, 104, 106, 109–110, 111; premotor, 24, 25, 51; motor, 48, 49–50, 51, 93, 103; speech production and, 93, 106; parasensory, 100; parietal, 103; temporal, 103
Creative thought, 1, 2, 4, 12
Crelin, Edmund, 64
Cruzon's syndrome, 60, 63
Culture: linguistics and, 36; Neanderthal, 65–67; chimpanzee, 66, 150–158,

176n1; language acquisition and, 134; transmission of, 149–150; early hominid, 158–164; altruism and, 169
Cummings, Jeffrey, 98
Curtiss, Susan, 120

Dalai Lama, 166–168, 170
Darwin, Charles, 3, 15, 53, 78, 111; natural selection theory, 2, 4–5, 7, 30, 80–81, 109, 152, 164, 165; on vocal tract, 54, 56
Darwinism, 4–5, 6–10; neural, 119
Deacon, Terrence, 25, 107, 179n15
Dementia, 94–95, 99, 178n9
Descartes, René, 4
Dialects, 154
Dibble, Howard, 160, 163
Distributed neural networks, 11, 136; computer-implemented studies, 30, 33–35, 116, 118, 123, 125, 175n5; analogy with electrical power systems, 31, 32–33; cognition and, 34–35; memory and, 104; word use and, 112–113, 115–119, 123; damaged, 122; language acquisition and, 139–140
DNA research, 172
Dogs, 112, 125
Dolphins, 175n1
Dopamine, 95–96
Down's syndrome, 119
Dresher, Elan, 181n3
Dysarthric speech, 93

Eastman, George, 9
Edelman, Gerald, 118, 175n5
Emotional responses, 78, 160
Engen, Elizabeth, 96
Engen, Trygg, 96
Essay on the Principle of Population (Malthus), 4–5
Evarts, Edward, 49, 50
Evolution (general discussion): of biological mechanisms, 2–3; hominid, 3, 80; preadaptation and, 6–8; adaptation and, 8–10; mammal, 16, 18–20, 21–22; logic of, 107; genetic variation and, 131–132
Exner, Sigmund, 28

Expression of the Emotions in Man and Animals, The (Darwin), 53

Fernald, Ann, 134
Filters in speech production, 41, 42, 43
Fish, swim bladders, 7, 53
Fodor, Jerry, 12
Formant frequency, 41–42, 44, 45, 57, 58, 77; of vowels, 59, 61–62, 65
Fossils: as evolutionary evidence, 18, 20, 109, 176n3; mammal, 20; Neanderthal, 63, 64, 65, 67, 68, 69, 163; hominid, 72, 162–163, 176n3; *Homo erectus*, 74; tool technology and, 160; burial rituals and, 162–164, 171; *Homo sapiens*, 171–172
Functional behavior, 8–10, 14
Functional branch-points, 8–10
Function vs. structure in evolution, 7–8
Fundamental frequency of phonation, 42

Gall, Franz, 13
Gardner, Allen, 139, 155
Gardner, Beatrix, 139, 155
General intelligence, 122–125
Genetics, 2, 6; genetic code, 4, 133; mutation and variation in, 4–5, 7, 131–132; primary reflexes and, 48
Geschwind, Norman, 15, 27, 78, 112, 178n8
Gleitman, Henry, 134, 182n6
Gleitman, Lila, 134, 146, 186n6
Goldstein, Kurt, 91–92, 96
Goodall, Jane, 52, 142, 150, 151, 152, 153, 157, 164
Gopnick, Alison, 145–146
Gould, Herbert, 60
Gracco, Vincent, 50
Groundings for the Metaphysics of Morals (Kant), 168

Hamilton, William, 165
Handedness, 79
Hayes, Cathy, 52
Hayes, Donald, 52
Hayes, Keith, 52
Hearing impairments, 96
Hebb, Donald, 30–31, 123, 124
Hebbian synaptic modification, 123–124

Hewes, Gordon, 106
Hierarchical categorization, 144–146
Hirsch-Pasek, Kathy, 146
Hollenbeck, Barbara, 168
Holloway, R. L., 176n3
Homo erectus, 74–76, 105, 160
Homo sapiens, 80, 84, 109, 160, 171–172
Howells, William, 63
Humboldt, Wilhelm von, 143

Ianco-Worall, Anita, 147
Imitation as cognitive strategy, 127, 141–142
Infant-mother communication, 13, 18, 20, 134–135
Information processing, 11, 28, 37–38, 81
Inner ear, 8
Intelligence, 20–21, 149; tests, 92; general, 122–125; linguistic dictionary and, 122–125
International Phonetic Alphabet, 40–41
Intonation, 3
Isolation cry, 3, 18–19, 21, 176n1

Jackson, Catherine, 120
Jakobson, Roman, 36–37
Jebel Qafzeh VI fossil, 76, 109–110, 160, 162–163
Jerison, Harry, 20
Johnson, Jacqueline, 138
Jurgens, Uwe, 106

Kagan, Jerome, 12, 135, 165
Kant, Immanuel, 168
Kebara Neanderthal fossil, 69
Keith, Arthur, 63–64
Kempler, Daniel, 119–120, 121
Kimura, Doreen, 79, 85, 86, 178n7
Klima, Edward, 90, 178nn7,10
KNM-ER-3733 fossil, 74, 76
Koch, Christof, 116
Kohonen, Tauvo, 34
Kuhl, Patricia, 46, 48

Labial stop consonants, 50–51, 58
La Chapelle-aux-Saints Neanderthal fossil, 64, 65, 67, 68
Landahl, Karen, 60

Landau, Barbara, 146
Language: evolution of, 1, 2–3, 72, 78, 112, 127; components, 2–3, 121, 148; biological mechanism of, 3; conceptualization, 3; creativeness of, 4, 81; organization, 12, 109; neural bases for, 13, 79–80, 98; aphasia and, 15, 93; brain mechanisms and, 15–16, 22, 24, 80, 84; neocortex and, 25–27; deficits, 34, 84, 86, 93; sounds, 37; thought and, 142–148; cognition and, 146, 161–162; comprehension, 146; learning and, 160; stress patterns of, 181n3. *See also* Speech; Syntax; Vocabulary
Language acquisition, 37; neural networks and, 3, 126; associative learning strategy, 127, 135, 138–140; imitation strategy, 127, 141–142; Universal Grammar and, 127–134; input method, 134–135; in ASL, 136–137; concept formation, 136–137; critical periods, 137–138; second, 138, 146–147; brain mechanisms and, 148; cognition and, 148
Larynx: speech production and, 3, 16, 27, 39–40, 41, 42; evolution of, 16, 53; function of, 54–56; location of, 64, 67, 69, 74
Lashley, Karl, 33
Lateralization theory of the brain, 79–80, 85, 104–105, 149
L-DOPA, 96
Learning: deficits, 24; distributed neural networks and, 30; associative, 31, 127, 135, 138–140; language and, 160
Lenneberg, Eric, 143
Lentiform nucleus, 101, 102
Limbic system, 16
Linguistic ability: cognition and, 10; creative thought and, 12; neural mechanisms and, 13; deficits, 15, 17, 24, 27; speech transmission rate and, 37–38; brain mechanisms and, 78, 125; general intelligence and, 122–125
Linguistic code, 38
Linguistic lexicon, 13
Lip rounding in speech, 83
Locationist model of the brain, 25, 28

Logic, formal, 114–115, 139
Lungs, 53–54, 108

MacLean, Paul, 16, 18
Malthus, Thomas, 4–5
Mammals, marine, 112
Mammary glands, 8, 20
Markowitsch, Hans, 100
Marler, P., 181n4
Marschack, Alexander, 161
Maynard-Smith, John, 165
Mayr, Ernest, 11
McCarthy, Rosaleen, 118
McCawley, James D., 180n2
Meltzoff, Andrew, 140, 141, 145–146
Memory, 11, 16, 30, 33, 82, 104
Mental retardation, 119
Mesulam, M.-Marsel, 28
Metter, Jeffrey, 94
Midbrain structures, 17, 21
Middle ear, 20
Miller, George, 37–38
Modular theory of brain structure, 12–14, 16
Mohanty, Ajit, 148
Mollusks, 123–124
Monkeys: separation calls, 18, 19, 21; brain size and structure, 22, 25, 26, 101, 102, 107; motor control, 24, 49–50; brain mechanisms, 45, 84, 86, 118, 179n15; speech perception, 46; vocalization, 53, 84–85, 106, 175n1, 177n3; altruistic behavior, 164
Morality, 1, 2, 9–10, 12, 172
Morphemes, 81, 87, 136, 144–145
Mosaic principle of living organisms, 6
Mother-child communication, 3, 18, 20, 134–135
Motor control: brain mechanisms and, 17, 21, 22, 25, 77; areas of the brain, 24, 25; neural mechanisms for, 25, 80; deficits, 27, 85, 96, 99, 119; manual, 29; vision and, 29–30; automatized, 50; voluntary, 72, 77; speech, 82–84, 85–86, 108, 109; subcortical disease and, 94; nonspeech, 96. *See also* Aphasia
Mouth structure, 20
Müller, Johannes, 39, 176n2
Mutism, 86

Naeser, Margaret, 93, 179n13
Nasalized speech sounds, 44, 51, 52
Nativist linguistic theory, 36–37, 133,
  136
Natural selection, 2, 30, 80–81; genetic
  variation and, 4–5, 7, 131; survival of
  fittest theory, 76–77; syntax and, 109;
  altruism and, 152, 164, 165
Neanderthal: fossils, 63, 64, 65, 67, 68,
  69; vocal tracts, 63–69, 110; culture,
  65–67; extinction, 76; brains, 110,
  176n3, 179n15; tool technology, 160
Negus, Victor, 53, 57, 63–64, 69
Neocortex. *See* Cortex: neocortex
Neural Darwinism, 119
Neural mechanisms: language acquisi-
  tion and, 3; linguistic ability and, 13,
  79, 125; motor control and, 25, 80; be-
  havior and, 28–29; structure of, 31;
  automatization and, 49–50; speech
  production and, 125
Neuroanatomy, 14
Neurology, 28
Neurophysiology, 84–91, 99–104
Neurotransmitters, 17
Newport, Elis, 134, 136, 137, 138
Nonnasal sounds, 57
Nottebohm, F., 181n4
Nursing mechanism, 20

Ojemann, George, 107
On the Origin of Species (Darwin), 2, 7,
  54
Operant conditioning, 139
Orofacial gestures, 51, 53

Palladium, 101
Palumbo, Carol, 93, 179n13
Parent, André, 178n12
Parkinson's disease (PD), 17, 94, 97–98,
  103
Pattnaik, Kabita, 148
Pavlov, Ivan, 124, 125
Periventricular white matter (PVWM),
  179n13
Peterson, Gordon, 61
Phonation, 39–40, 42, 45, 86, 134
Phonetic segments, 37
Photographic technology, 8–9

Phrenology, 12, 23, 25
Physiology, 14
Play as cognitive strategy, 127, 140
Poisner, Howard, 90, 178nn7,10
Population theory, 4–5
Positron emission tomography (PET),
  92–94, 95, 99
Preadaptation, 6–8, 53, 166, 176n1; of
  speech mechanisms, 56, 86; Universal
  Grammar and, 108–109
Premack, David, 157, 161
Premotor areas of the brain, 24, 25, 51
Progressive supranuclear palsy (PSP),
  94, 95, 99
Punctuated equilibria, 6–7
Putamen, 22, 87, 101, 103
Putnam, Hilary, 115

Quantal speech sounds, 57–58, 74

Radar and modular design, 14
Radke-Yarrow, Marian, 168
Redundancy, 107
Reinforcement concept, 138–139
Religious thought in early hominid cul-
  ture, 161–164, 172
Reptilian complex, 16, 21, 22
Respiration, 74
Rhode Island Test of Language Struc-
  ture (RITLS), 96–98
Right-ear advantage, 105
Rodents, 101, 102–103, 139

Sapir, Edward, 143
Sapir-Whorf hypothesis, 142–144
Savage-Rumbaugh, Sue, 155–156
Sejnowski, Terrance, 35, 116
Selfless behavior. *See* Altruism
Separation calls. *See* Isolation calls
Shanidar (Iraq), 160, 163
Sign language, 1, 38, 90–91, 105–107.
  See also American Sign Language
  (ASL)
Singleton, Jenny, 136, 137
Skhul V fossil, 76, 109–110, 160, 163
Skinner, B. F., 138–139, 140
Snow, Catherine, 134–135, 181n4
Social organization of chimpanzees,
  151–152, 157

Spearman, Carl, 122
Speech: brain mechanisms and, 1, 3–4, 36, 45–46, 72, 111; evolution of, 1–2, 3, 37, 39, 63; communication speed and, 3; voluntary, 21, 105–107, 109, 111; comprehension deficits, 27, 87, 96–98, 119–122; biological mechanisms for, 37, 39, 57; linguistics and, 37; Neanderthal, 37, 67; sounds, 37–39, 44–46, 47, 57–58, 105; transmission rate, 37–38, 42, 45, 58, 59, 74; physiology of, 39–42; nasalized, 44, 65; decoding, 45–46, 77; selective advantages of, 57–59; encoding, 58, 59, 77, 78, 82; vocal tract evolution and, 60–63; hominid, 70; anatomical specialization for, 72, 75; origins of, 72–77; motor control and, 82–84
Speech perception: brain mechanisms and, 13, 45–46; filter function of vocal tract in, 42–44; nasalized speech sounds, 44, 51, 52, 57; decoding, 45–46; detectors, 45; genetically transmitted ability for, 45; acoustic cues in, 46–47; vocal tract normalization and, 46–48; brain lateralization and, 105, 179n14; motor theory of, 179n14
Speech production: brain mechanisms and, 1, 3–4, 17, 20, 24, 36, 45–46, 72, 74, 79, 82–83, 85–86, 106–107, 142; evolution of, 1, 2, 3; motor control of, 3–4, 21, 83, 85; deficits, 17, 27, 86, 87, 93, 98, 103, 119–122; anatomical structure and, 20, 21, 39–41, 56, 60; voluntary, 21, 105–107, 109, 111; filters, 41, 42, 43; automatization and, 48–51; of nonnasal sounds, 57; source-filter theory of, 176n2; lateralized, 179n14. *See also* Aphasia
Spurzheim, Johann, 13
Sternberg, Robert, 123
Stevens, Kenneth, 58–59
Stress patterns of language, 181n3
Stuss, Donald, 85, 87, 99–100, 177n5
Subcortical diseases, 93–95, 98
Substantia nigra, 95, 101, 103
Subthalamic nucleus, 101
Suppala, Ted, 137, 138
Supplementary motor association area (SMA), 106

Supralaryngeal vocal tract, 39, 41, 42, 44, 47, 72; evolution of, 53–72; location of, 54–56; nonhuman, 55; Neanderthal, 63–69, 73; hominid, 73
Sutton, Dwight, 106
Swim bladders, 7, 53
Synapses, 30, 31–32, 124, 175n5
Syntax, 177n2; evolution of, 1–2, 3, 80, 104–110; communication speed and, 3, 82; rules, 4, 34, 81, 83, 104, 128; brain mechanisms and, 12, 78, 111, 125, 142; comprehension deficits and, 27; encoding, 82; as species-specific phenomenon, 82; deficits, 87, 91, 93, 96–98, 103, 119, 121; aphasia and, 89; on-line comprehension of, 90; variation in, 109; Universal Grammar and, 127–128; acquisition, 128–129; in ASL, 136, 157; cognition and, 146

Teeth, 56
Thorndike, Edward L., 122
Thought: creative, 1, 2, 4, 12; evolution of, 2, 104–110; brain mechanisms and, 21, 22, 80, 111; encoding, 82; abstract, 109, 110; defined, 116; animal, 125; language and, 142–148; religious, 161–164
Tibetan Buddhist theology, 166–168
Tomasello, Michael, 135
Tongue, 54, 58–59, 67
Tool use, 104–106, 149, 182n2; chimpanzee, 104, 105, 142, 150–151, 159; hominid, 158–161
Torture, 9–10
Tracer techniques in neurophysiology, 102
Trinkhaus, Eric, 63

Udwin, O., 121
Universal Grammar (UG), 6, 108–109, 127–134, 180nn1,2, 182n6; ASL and, 136–137

Verb tense, 116–117, 132
Vigilance deficits, 24
Vision, 6; stabilizing, 29–30; color, 58, 143
Visual association area, 24
Vocabulary, 3, 12, 112, 113, 145

Vocal cords, 4
Vocalization patterns, 18–20; brain mechanisms and, 21, 105; voluntary, 21, 72, 77, 105–107; species-specific, 45; of monkeys and apes, 52, 53, 62, 72, 84–85, 105, 106, 175n1, 177n3
Vocal tract, 39, 41–42; normalization, 46–48, 58, 77; primate, 58; evolution of, 60–63, 109, 110; phonetic capacity, 62; Neanderthal, 63–69; reconstruction of, 69–72; nonhuman, 70–72, 74; control of, 72, 74. *See also* Supralaryngeal vocal tract
Voice-onset time (VOT), 45–46, 86, 97–98
Vowels and vowel sounds, 41, 42–44, 50; nasalized, 51; misidentification of, 58, 60; formant frequency of, 59, 61–62, 65; rounded, 83–84; infant identification of, 137

Warlpiri language, 177n2
Warrington, E. K., 118
Waxman, Sandra, 144–145
Wechsler, David, 122
Wechsler Test of Adult Intelligence, 123
Wernicke, Carl, 15, 121
Wernicke's aphasia, 27, 34, 90, 91, 94, 178n8; brain's dictionary and, 121–122
Wernicke's area, 23, 27, 34, 90
Whorf, Benjamin, 143
William's syndrome, 120–121
Woodruff, Guy, 157
Word: acquisition and interpretation, 112; use, 112, 115–119; meaning, 113–115, 116
Wulfeck, Beverly, 90

Zahn-Wexler, Carolyn, 168–169

∿0